SYSTEMS ENGINEERING and SAFETY

BUILDING THE BRIDGE

Peter J. Glismann

SYSTEMS ENGINEERING and SAFETY

BUILDING THE BRIDGE

CRC Press
Taylor & Francis Group
Boca Raton London New York

CRC Press is an imprint of the
Taylor & Francis Group, an **informa** business

CRC Press
Taylor & Francis Group
6000 Broken Sound Parkway NW, Suite 300
Boca Raton, FL 33487-2742

© 2013 by Taylor & Francis Group, LLC
CRC Press is an imprint of Taylor & Francis Group, an Informa business

No claim to original U.S. Government works

Printed on acid-free paper
Version Date: 20130403

International Standard Book Number-13: 978-1-4665-5212-8 (Paperback)

Library of Congress Cataloging-in-Publication Data

Glismann, Peter J.
    Systems engineering and safety : building the bridge / author, Peter J. Glismann.
    pages cm
    Summary: "Systems engineering principles are currently being applied to system safety best practices in several industries, as well as state and local governments. This book covers the payoff in both dollars and goodwill to the investment made in merging those two important and often neglected disciplines. It can be read, understood, and acted upon by the Chief Executive Officer of a multinational corporation, right down to the line manager of systems engineering or the technical professional in the safety department. The value in terms of cost savings, be it in the form of financial, human, or social capital, is clearly illustrated with real world examples"-- Provided by publisher.
    Includes bibliographical references and index.
    ISBN 978-1-4665-5212-8 (pbk.)
    1. Reliability (Engineering) 2. Safety factor in engineering. 3. Systems engineering. 4. System failures (Engineering)--Prevention. I. Title.

    TA169.G54 2013
    620'.0042--dc23   2013008811

Visit the Taylor & Francis Web site at
http://www.taylorandfrancis.com

and the CRC Press Web site at
http://www.crcpress.com

This book is dedicated to my father, Peter P. Glismann: July 16, 1932—March 24, 2012. His valiant battle with cancer was fought with the same blend of intense work ethic and charm that he always brought to all other endeavors in his life. He will always be an inspiration to me.

Contents

Preface

This book is for systems engineering and safety professionals in all disciplines and industries, worldwide. If your knowledge of one discipline is aware of the existence and value of the other, then this book will enhance that awareness. In the event that you are involved in one discipline without knowledge of the other, then this book will help you, from the point of view of another equally important professional discipline, see how a relationship will benefit your organization and industry. If the function you serve is not directly involved with either systems engineering or safety, then I believe that this book will, in a clear and concise way, enlighten you to the value of each discipline and, even more importantly, illustrate the need to build a bridge between the two disciplines.

This book is unique because it was written by someone (me) who left college without an understanding, from either an educational or professional point of view, of systems engineering or safety. Over the course of the past twenty-five years, I have been enlightened, through continuing education and experience, while serving as both a systems engineer and a safety subject matter expert, and I have realized the need for integration and synergy between these two disciplines.

This book describes and gives examples of systems engineering methodology and safety tools, and describes the management needed to build the bridge between these two disciplines. This bridge must be properly built at the outset and maintained throughout the life cycle of the project.

The strength of this book is that it can be read, understood, and hopefully acted upon by the chief executive officer of a corporation, right down to the line manager of systems engineering or the subject matter expert in the safety department. This value can be measured in cost savings, be it in the form of human, social, or financial capital. Real-world examples will illustrate this point.

The challenge ahead is for all of us to reach across professional disciplines that we may have overlooked in the past, to work together on an interpersonal level, and to build the bridge between systems engineering and safety that will enlighten and help us. We need to overcome inertia and act on lessons learned. To all the systems engineering and safety professionals who utilize this book, please enjoy reading and keep up the good work.

Acknowledgments

I must first acknowledge my editor, Cindy Carelli at CRC Press. Thank you for seeing the potential of this idea, and for guiding me through the process of making this effort a reality.

I am thankful for the love and support of my new bride, Jacquelyn. She has been my staunchest supporter and my fiercest critic throughout the creation of this book.

I owe a debt of gratitude to my teachers at every step in my educational development, who taught me at Monsignor Farrell High School, The Ohio State University, and The George Washington University. They recognized that my academic ambition far exceeded my level of effort, and they encouraged me to build the bridge between that ambition and effort, at each and every step.

I must thank the great engineers and technicians with whom I worked at Naval Surface Warfare Center, Indian Head, Maryland, where I began my career more than twenty-five years ago. It was there that I first learned the meaning of having a strong work ethic and professionalism, and the first place where I understood the need for synergy between systems engineering and safety.

I thank my "murder board" at the Metropolitan Transportation Authority, New York City Transit: Tom Lamb, Chief, Innovation and Technology; and John Soucheck, director of Bus and Rail Field Operations. This book would not have evolved from an idea to a technical presentation for the National Safety Council Congress & Expo to its current state, if not for their clever, enlightened insight and constructive criticism every step of the way.

Finally, I must acknowledge my parents, Peter P. and Margaret M. Glismann. As a man with a Professional Engineer's license in mechanical engineering, accreditation as a Certified Safety Professional, and a Master of Science degree in systems engineering and management, this book is yet another attempt on my part to do justice to the high hopes of accomplishment that my parents always held for me.

About the Author

Peter J. Glismann is a Certified Safety Professional and a licensed Professional Engineer. He has served the defense, construction, supply chain management, and transportation industries as a consultant in the areas of systems engineering, system safety, program management, and special inspection. In addition, he is a member of the National Speakers Association, a Master of Special Inspection, and has supported clients as a forensics engineer and expert witness.

Introduction

This story starts in December of 1985. I took an early end to my winter break and left my home in New York in order to get to Columbus, Ohio, so that I could "road trip" with my college roommates to Orlando, Florida, to watch our beloved football team, The Ohio State University Buckeyes, play against the Brigham Young University Cougars in the Citrus Bowl. While in Florida, we made what, as undergraduate engineering majors, we called a "pilgrimage" to the National Aeronautics and Space Administration (NASA) Kennedy Space Flight Center (also known as Cape Canaveral) to tour the facility and to see the marvel of space exploration, a Space Shuttle. As luck would have it, both the Shuttle *Challenger* and the Shuttle *Columbia*, both soon to be launched, were on their respective launch pads during our tour. As a student in the aeronautical and astronautical engineering program, I remember how excited I felt to see engineering at its finest, in all its vast glory. Soon both shuttles would travel to space, performing experiments and setting a precedent for further expansion of an exciting industry. I remember the confidence I felt on that trip, knowing that NASA had applied lessons learned during the Apollo program, in which three astronauts died in a launch accident in 1967, only for the program to reemerge in a safer and redesigned state, and eventually land a man on the Moon and bring him home safely, as President John F. Kennedy had predicted nearly a decade before. As the photo on the cover of this book shows, there they were—the space shuttles, the vehicles that would bring a safe and effective form of space exploration to an eager and excited nation.

A few weeks later, I was back at school and working part-time as a research assistant on January 28, 1986. A senior technician came into the office and informed us that the Space Shuttle *Challenger* exploded, killing all seven astronauts aboard. I was overwhelmed with shock, anger, grief, and a sense of disbelief that the awesome display I had witnessed ended in disaster. How could this happen? What could possibly have gone wrong? The months and years that followed identified the culprit: faulty O-ring seals in the solid rocket booster solid rocket motors, which, after a redesign, allowed further shuttle flights. As I followed the story in the press and all the technical journals available, I read that lessons were learned, the culture at NASA changed, and things would begin anew. As did the Apollo program, the shuttle fleet survived disaster to fly yet again—in a safer and more efficient state, for sure.

After my graduation, my career saw me work in propellant and explosive manufacturing, missile design, project management, systems integration and test, systems engineering, and system safety. The "tech boom" of the late 1990s presented an opportunity to move back to New York and transfer my skill set from the defense industry into information technology support for the finance, supply chain, and transportation industries. Every step of the way, I learned how safety would become a more important part of the work in which I was involved. In the Department of Defense, it was the safety engineers who had the authority to shut down a manufacturing process or cancel a test. I learned early on in my career how to perform the safety expert's job better than the safety expert could. Processes had to be kept online, and tests had to be performed. I quickly learned to incorporate safety into my skill set in order to do my job well.

While working as a civilian aerospace engineer at the Naval Surface Warfare Center, Indian Head, Maryland, I went to night school to earn my master's degree in systems engineering and management. While I worked in both manufacturing and design, I learned the synergy between systems engineering and safety, and I applied those skills in the field, at live fire missile exercises and tests, first article inspections, and systems integration activities nationwide. I then had the opportunity to apply my advanced-degree systems engineering skill set at TRW.

My career then took me to New York City, where I now work as manager of New Technology Application Safety at the Metropolitan Transportation Authority, New York City Transit. I was hired because I had a résumé with two separate but very important skills: systems engineering and safety. On February 1, 2003, I was doing my laundry at a corner laundromat on the Upper East Side. It had a few televisions set to mute, so it took me a minute to see the news: the Space Shuttle *Columbia* flight ended in disaster, killing all seven astronauts, as had the Space Shuttle *Challenger* more than seventeen years prior. I thought back to those previous memories of excitement in December 1985 and shock in January 1986. I realized on that morning in 2003 that I had not just a professional interest in space flight, but also a personal connection. I went from seeing with my own eyes both vehicles in a state of proud display, to hearing the news that both vehicles met a tragic end, due to two separate, yet (arguably) preventable breaches of safety. The systems engineering process had failed, safety had been compromised, and lives had been lost.

As the global economy expands and more advanced technology continues to be utilized worldwide, projects and processes will continue to become more complex. A paradigm will be essential to manage this vast expansion. The bridge must be built between systems engineering and safety; and these two separate but very important disciplines must be integrated. Therein lies the purpose of this book.

Chapter 1 lays out the purpose of the methodology of systems engineering and the tools of safety. The main building block of this bridge is management, and the culture, commitment, communication, and coordination that management must provide. Chapter 2 describes systems engineering methodology: the life cycle, processes, and management. Chapter 3 describes the tool set of safety: techniques, processes, and management. Chapter 4 discusses the technical processes with which systems engineers and safety professionals must be familiar. Chapter 5 merges management, systems engineering, and safety into the life cycle through project processes. Chapter 6 examines the roles and responsibilities of management, and provides a breakdown theory of safety in the management processes: The Glismann Effect. Chapter 7 uses real-world examples—the explosion aboard the battleship USS *Iowa*, where forty-seven sailors were killed in 1989, and the previously mentioned NASA Shuttle disasters: *Challenger* in 1986 and *Columbia* in 2003. Chapter 8 discusses the road ahead, with a plea for all stakeholders involved to do their part to embrace systems engineering and safety, and to build the bridge.

1 Scope

1.1 PURPOSE OF SYSTEMS ENGINEERING

The purpose of systems engineering is to integrate parts, which make up a whole system, by applying knowledge of technical principles (i.e., the practice of engineering) and ensure that those parts function properly. By way of definition:

- A *system* is a set of things working together as a complex whole; a group of related hardware units or programs, or both.
- *Engineering* is a field of study concerned with modification or development in a particular area; the branch of science and technology concerned with the design, building, and use of engines, machines, and structures.
- *Systems engineering* is the interdisciplinary approach governing the total technical and managerial effort required to transform a set of customer needs, expectations, and constraints into a solution, and to support that solution throughout its life cycle.
- *Life-cycle model:* A partitioning of the life of a product, service, project, workgroup, or set of work activities into phases.

Systems engineering has been defined, practiced, and studied dating back to the 1940s. The first instance of the words "systems engineering" in the *New York Times Archive* dates back to an item from November 1, 1953: "A new course in systems engineering, geared to the complex requirements of expanding military and civilian developments in electronics and science, has been inaugurated at the University of Pennsylvania's Moore School of Electrical Engineering." A cursory search of Internet articles and lectures would have you believe that systems engineering started with Watt's Flyball Governor in 1788. Because of his work in the Army Air Corps Statistical Control during World War II, followed by stints as president of the Ford Motor Company and Secretary of Defense during the Vietnam era, Robert S. McNamara has been recognized by some systems engineers as the father of the discipline. Regardless of the exact origins, the intent to achieve a goal by assembling parts into a system and applying technical knowledge to efficiently utilize that system has always existed.

1.2 PURPOSE OF SAFETY

Safety is defined as the condition of being protected from, or unlikely to cause danger, risk, or injury. In the context of applying systems engineering throughout a project lifecycle, then denoting something designed to prevent injury or damage, the purpose of safety is to establish the fact that a condition in a system will not cause danger, risk, or injury. Safety, also known as System Safety, as a field of expertise for a subject matter expert (SME) such as a safety engineer, is concerned with two primary issues: hazards and failures. As a subset to failures, a third concern is faults. By way of definition,

- A *hazard* is a system state or set of conditions that, together with a particular set of worst-case environmental conditions (Probability and Severity, as discussed in Chapter 3), will lead to an accident or loss of system or system function.
- A *failure* is an improper condition that prevents a system from delivering the expected performance of that system to the user. A failure may require unscheduled equipment maintenance or replacement to restore the affected equipment to its normal operating condition.
- A *fault* is a method of determining cause or affixing blame. It is best safety practice to define faults in terms of failures. Although not every fault in a system causes a failure in that system, faults that either exist within or interact with a system may result in a system failure. When a system is determined to have faults that do not cause failures in the system, the system is said to be "fault tolerant."
- *Risk* is defined as an expression of possible loss over a specific period of time or number of operational cycles. It may be expressed as the product of hazard severity and hazard probability.

Faults are analyzed through a graphical representation of causality known as Fault Tree Analysis (FTA). Faults are used to analyze the effect of failures on the system, subsystem, or operating environment (i.e., to facilities, equipment, or personnel). Failures are associated with a quantitative analysis of the design of the system. Hazards are assessed qualitatively, and must be analyzed and either eliminated or reduced to an acceptable level of risk through a mitigation process. The relationship between faults, failures, and hazards may best be understood as follows: not all faults are failures and not all failures present a hazard to the system.

The mapping of safety tasks to the system life cycle becomes obvious as cycles move from design to construction or development and use. Before construction of physical facilities or development of products and processes begins, failures must be analyzed to determine if any redesign is required early in the life cycle.

As systems move into the hands of the user community, hazards must be controlled, either by changes to operating procedures or other methods, so that the level of risk is either eliminated or found to have decreased to an acceptable threshold throughout the life of the system. A robust safety presence must be integrated into the systems engineering process during the entire life cycle.

1.3   NEED TO BUILD THE BRIDGE

Often, circular definitions have, incorrectly, led the words "hazard" and "failure" to be used interchangeably. Also, in many occurrences, hazard mitigation has been forced on a project because failure analysis did not identify opportunities for redesign early enough in the life cycle to leave project stakeholders with any other choice

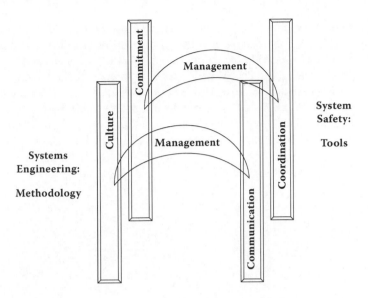

FIGURE 1.1 M – T – M.

than to resolve risk via operational procedures, rather than through changes to the design. The tools and processes of safety must be incorporated early in the systems engineering methodology, and the resources committed must continue to exist as a valuable part of the development and decision-making team throughout.

As shown in Figure 1.1, the key building blocks of the bridge between the methodology of systems engineering and the tools of safety can be defined by the acronym M – T – M, which stands for

Methodology – Tools – Management

Figure 1.2 illustrates the overall relationship. Systems engineering provides the methodology; Safety provides the tools of identification, analysis, and mitigation;

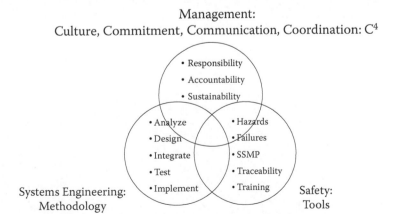

FIGURE 1.2 Management: Culture, Commitment, Communication, and Coordination: C^4.

and Management provides what will be referred to as C^4: the culture, commitment, communication, and coordination. It is the responsibility of Management to instill a culture of safety; to provide the commitment, by way of resources and enforcement; to communicate the decisions with respect to safety made at the highest levels throughout the organization; and to provide coordination between different and often competitive resources within an organization to achieve safe project success. Safe, successful organizations have this type of management. Chapter 7 provides examples of what has happened when that Management building block fails in its effort.

2 Systems Engineering *Methodology*

2.1 SYSTEMS ENGINEERING LIFE CYCLE

The time frame in which a solution to a customer's needs is obtained through the systems engineering methodology is known as the life cycle. Also referred to as the System Life Cycle, or the System Development Life Cycle (SDLC), the purpose in defining the system life cycle is to establish a framework for meeting the customer's needs in an orderly, efficient, and safe manner. As we shall see in Chapter 3, safety tools should be integrated into the life cycle early and throughout in order to achieve this goal.

The International Council on Systems Engineering (INCOSE) *Systems Engineering Handbook* defines the steps of a life cycle as follows:

- *Concept stage:* In the Concept stage, the requirements of the stakeholders are identified, clarified, and documented. Subsystem elements of the overall system are generated. Hardware and software modeling are used in order to establish the requirements, and begin to validate the system design. Safety objectives are documented and the Preliminary Hazard Analysis (PHA) is performed.
- *Development stage:* In the Development stage, integration of subsystem elements, between each other and the overall system, and validation and verification (V&V) activities, including proof of safety concepts, begin. The System Hazard Analysis (SHA) is generated. The system baseline is further defined and documented as these activities progress.
- *Production stage:* The Production stage is where the first article of a hardware item or the beta-version of software is produced. The Subsystem Hazard Analysis (SSHA) is generated. Scale-up is done in order to resolve production problems and costs, and to refine system capabilities. Any changes to initial system requirements or revalidation efforts must be accomplished in the Production stage.
- *Utilization/Support stage:* This is the stage where the system is operated as intended, and delivers the solution to the stakeholder's requirements. Upgrades to the system intended to enhance system capabilities may be introduced during the Utilization stage, as long as the systems engineers ensure that the safety baseline is maintained, and mishaps are identified and corrected according to the Operating and Support Hazard Analysis (O&SHA).

- *Retirement stage:* The Retirement stage is where the operating system is removed from service. Safe disposal of equipment, facilities, and waste must be ensured. The "grave" portion of the term "cradle to grave" became essential in the latter half of the twentieth century, as environmental problems mandated the creation of U.S. Government programs such as "Superfund." In modern systems engineering methodology, system retirement requirements are considered during the Development stage.

The integration of safety into the system life-cycle stages is shown in Figure 2.1. The choice of steps within the life cycle may be redefined as project requirements mature. Modification of the life cycle for specific customer or program needs is known as "tailoring."

To develop the life cycle, systems engineers must consider the business case, the budget aspect (funding), and the technical aspect (product). Project management enters the effort in the development of decision gates, also known as Milestones or Reviews. Milestones are represented as major events throughout the life cycle. In Chapter 5, we review the integration of safety into milestones. Reviews occur at specified times during a project or program contract and, by convention, usually occur at the Preliminary, Critical, and Final stages of system design.

Various life-cycle models may be utilized in defining the start, stop, and process activities appropriate to the life-cycle stages. Figure 2.2 shows three of the "best practice" life-cycle models used in systems engineering. These are known as

- *Waterfall Model:* This model may be referred to as the "throw it over the fence" model. Because the stages of this model have so often been utilized, it may be employed by both veteran and newer stakeholders, and between technical and back-office personnel. This model is best employed when requirements are clear and architecture or infrastructure is understood. Additional steps and feedback loops between steps may be added in the tailoring process.

Proactive Safety Effort				Reactive Safety Effort			
Concept Retirement Stage		**Development Stage**		**Production Stage**		**Utilization Stage**	
System Safety Management Plan (SSMP)	Preliminary Hazard Analysis (PHA)	Subsystem Hazard Analysis (SHA)	Subsystem Hazard Analysis (SSHA)	Operating & Support Hazard Analysis (O&SHA)	Maintenance of Safety Baseline	Mishap Investigation and Correction	Safe Disposal
	Preliminary Design Review (PDR)	Critical Design Review (CDR)		Final Design Review (FDR)			

FIGURE 2.1   Safety in the system life cycle.

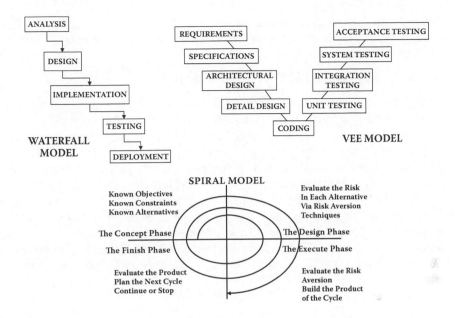

FIGURE 2.2 Life-cycle models.

- *"Vee" model:* This model is used to visualize the systems engineering efforts. It highlights the need to define verification plans during requirements development, the need for continuous validation with stakeholders, and the importance of continuous risk and opportunity assessment. The left side of the Vee shows the evolution of the system baseline from requirements to detailed design. The right side of the Vee illustrates the evolution of system testing.
- *Spiral Model:* Also known as "build a little, test a little," the Spiral model is used mostly for software development. Use of this model requires stakeholders with software subject matter expertise, and management commitment to keep knowledgeable personnel o the development team throughout the iterative steps of project execution.

Just as in choosing the steps of the life cycle, the models used may be influenced by the incorporation of safety tools. Feedback loops are essential in the life-cycle model, so that issues involving cost, schedule, and safety may be addressed before the project matures beyond the ability of systems engineers and safety professionals to successfully integrate their respective disciplines throughout the management of the project.

2.2 SYSTEMS ENGINEERING PROCESS

The International Standards Organization document ISO/IEC 15288:2008 System and Software Engineering, System Lifecycle Processes defines the Technical Processes of the life cycle, discussed in Chapter 4, and the Project Processes, discussed in

Chapter 5. These chapters discuss the integration of safety into the management of the life cycle throughout the Technical Processes and the Project Processes.

2.3 SYSTEMS ENGINEERING MANAGEMENT

The definitive management document for the systems engineering process is the Systems Engineering Management Plan, or SEMP, as described in the book *Systems Engineering Management* by Benjamin Blanchard. Generation of the SEMP begins in the Concept stage. The SEMP is the highest engineering planning document in the document tree hierarchy. It covers all management functions associated with the performance of systems engineering activities for a given program. Figure 2.3 shows an example SEMP outline. Just as with the life-cycle model, the SEMP may be tailored to suit the size and complexity of the program or project undertaken to fulfill the customer's needs. Technical and Project Processes can be managed throughout the plan described in the SEMP. As we shall see in Chapter 3, safety management, as achieved through the System Safety Management Plan, or SSMP, is an offshoot document of the SEMP.

1.0 Overview

2.0 Purpose and Scope

3.0 Application(s)

4.0 Technical Program Planning, Implementation, and Control (Part I)
    4.1 Introduction
    4.2 Program Requirements/Statement of Work
    4.3 Organizational Information
        4.3.1 Roles and Responsibilities
        4.3.2 Interrelationships
        4.3.3 Communication Methodology
        4.3.4 Customer Relationship Management
        4.3.5 Supplier Relationship Control
    4.4 Program Planning
    4.5 Technical Performance Measurement
    4.6 Program Cost Reporting and Control
    4.7 Design Reviews
        4.7.1 Preliminary Design Review
        4.7.2 Critical Design Review
        4.7.3 Final Design Review
    4.8 Technical Audits
    4.9 Configuration Management
    4.10 Risk Management

5.0 System Engineering Process (Part II)
    5.1 Needs Analysis and Feasibility Study
    5.2 Operational Requirements
    5.3 Maintenance Philosophy
    5.4 Functional Analysis
    5.5 Requirements Allocation
    5.6 System Synthesis, Analysis, and Trade-Offs
    5.7 Design Integration and Support
    5.8 System Test and Evaluation
    5.9 Production/Construction Support
    5.10 System Modification: Change Management Implementation and Control

6.0 Engineering Specialty Integration: Roles and Responsibilities
    6.1 Reliability Engineering
    6.2 Quality Engineering: Assurance and Control
    6.3 Safety Management

7.0 References
    7.1 Specifications
    7.2 Standards
    7.3 Plans

FIGURE 2.3 Sample Systems Engineering Management Plan (SEMP).

3 Safety
Tools

3.1 SAFETY TECHNIQUES

Safety techniques fall into two categories: Hazard Analysis (HA) and Failure Analysis (FA). Faults, analyzed in the Fault Tree Analysis (FTA) technique, are tied to failure analysis through a coordinated specialty engineering effort. FA is performed using the Failure Modes and Effects Analysis (FMEA) and Failure Modes and Effects Criticality Analysis (FMECA) techniques. FA is a specialty engineering responsibility of reliability engineering. Just as the integration of systems engineering and safety is done in the SEMP, roles and responsibilities of reliability engineering, quality assurance and quality control, and other areas of respective specialty engineering that must integrate into safety management of the system, must be integrated throughout the life cycle in the System Safety Management Plan (SSMP), known in some industries as the Product Safety Plan (PSP). Figure 3.1 shows a sample SSMP. Safety Management is discussed in Section 3.3.

The HA safety technique is described in MIL-STD-882C: Military Standard System Safety Program Requirements. This standard provides uniform requirements in a system safety program to identify hazards of a system and to impose design requirements and management controls to prevent mishaps. Early hazard identification and elimination or reduction of associated risk are essential to a formal safety program.

All industries follow a similar linear process to identify and analyze system hazards. The Hazard Resolution Process in the transportation industry, as described in the *US Department of Transportation Federal Transportation Administration Hazard Analysis Guidelines*, is as follows:

- Define the system
- Identify hazards
- Assess hazards
- Resolve hazards
- Follow up

HA is qualitatively performed in both narrative and matrix formats. The common matrix used across most disciplines uses Categories of Severity and Levels of Probability. Hazard Severity Categories from I to IV provide a measure of the worst credible mishap resulting from personnel error, environmental conditions, design

1.0   Overview
    1.1   Roles and Responsibilities

2.0   Internal Safety Management
    2.1   System Safety Working Group
    2.2   System Safety Certification Board

3.0   External Safety Management
    3.1   Consultant Activities
    3.2   Vendor Activities
    3.3   Subcontractor Activities

4.0   Safety Reviews
    4.1   Preliminary Hazard Analysis
    4.2   System Hazard Analysis
    4.3   Subsystem Hazard Analysis
    4.4   Operating and Support Hazard Analysis

5.0   Safety Test and Evaluation
    5.1   Proof of Concept
    5.2   Proof of Safety Case
    5.3   Beneficial Use
    5.4   Final Safety Certification

6.0   Safety Reporting
    6.1   Safety Requirements
    6.2   Hazard Generation
    6.3   Failure Analysis
    6.4   Hazard Log
    6.5   Final Safety Report

7.0   Safety Training
    7.1   Operating Personnel Training
    7.2   Maintenance & Support Personnel Training

8.0   Safety Audits
    8.1   Internal Audits
    8.2   External Audits: Vendors & Subcontractors
    8.3   Corrective Action

FIGURE 3.1   Sample Systems Safety Management Plan.

inadequacies, procedural deficiencies, and failures at the system, subsystem, or component level, as identified in the FTA, FMEA, or FMECA, as follows:

    Hazard Severity Categories:
        Category I: Catastrophic: Death, system loss or severe environmental damage.
        Category II: Critical: Severe injury, severe occupational illness, major system or environmental damage.

Category III: Marginal: Minor injury, minor occupational illness, minor system or environmental damage.

Category IV: Negligible. Less than minor injury, occupational illness, or less than system or environmental damage.

Hazard Probability Levels are qualitatively ranked from A to E. An example description from the Federal Transportation Administration follows:

(A) Frequent: Likely to occur frequently. Mean Time Between Events (MTBE) is less than 1,000 operating hours.
(B) Probable: Will occur several times during the life of an item. MTBE is equal to or greater than 1,000 operating hours and less than 100,000 operating hours.
(C) Occasional: Likely to occur sometime in the life of an item. MTBE is equal to or greater than 100,000 operating hours and less than 1,000,000 operating hours.
(D) Remote: Unlikely but possible to occur during the life of an item. MTBE is greater than 1,000,000 operating hours and less than 100,000,000 operating hours.
(E) Improbable: So unlikely, it can be assumed occurrence may not be experienced. MTBE is greater than 100,000,000 operating hours.

Table 3.1 is a representative matrix that combines hazard severity and hazard probability, so that risk assessment criteria may be applied to determine the acceptance of the risk, or to identify the need for corrective action by means of redesign, procedural instructions, or other means, to eliminate or reduce the risk to an acceptable level.

TABLE 3.1
Example Hazard Risk Assessment Matrix

	Severity			
	(I)	**(II)**	**(III)**	**(IV)**
Probability	**Catastrophic**	**Critical**	**Marginal**	**Negligible**
(A) Frequent	*IA*	*IIA*	*IIIA*	IVA
(B) Probable	*IB*	*IIB*	**IIIB**	IVB
(C) Occasional	*IC*	IIC	**IIIC**	IVC
(D) Remote	**ID**	**IID**	IIID	IVD
(E) Improbable	IE	IIE	IIIE	IVE

Hazard Risk Index	Suggested Criteria
IA, IB, IC, IIA, IIB, IIIA	*Unacceptable*
ID, IIC, IID, IIIB, IIIC	**Undesirable**
IE, IIE, IIID, IIIE, IVA, IVB	Acceptable with review
IVC, IVD, IVE	Acceptable without review

Hazards are captured in life-cycle milestones and tracked in the Hazard Log, which is a document that contains resolution and mitigation of hazards to an acceptable level of risk captured in the Hazard Mitigation Forms. Safety processes are discussed in Section 3.2.

The FA safety technique is described in MIL-STD-1629A: Procedures for Performing a Failure Mode, Effects and Criticality Analysis. FA is performed to provide an iterative safety decision tool to evaluate the feasibility and adequacy of the design approach. By way of linear explanation, new technology systems define and analyze hazards before failure analysis is performed, because historical design failure data do not exist. The earlier the quantitative results can be applied to a system, the better the information available to stakeholders to perform cost and schedule trade-offs. For system upgrades to legacy systems (i.e., systems with solid historical performance and reliability data available), hazards and failures can be analyzed concurrently early in the system life cycle. Failure modes and causes begin with an examination of the following typical failure conditions:

- Premature operation
- Failure to operate at a prescribed time
- Intermittent operation
- Failure to cease operation at a prescribed time
- Loss of output or failure during operation
- Degraded output or operational capability
- Other unique failure conditions, as applicable, based on system characteristics and other operational constraints

The FMEA process is as follows:

- Define the system to be analyzed. A System Design Document should provide a complete system definition, which includes identification of internal and interface functions, expected performance at all system and subsystem indenture levels, system restraints, failure definitions, operational tasks, environmental profiles, equipment utilization, and the functions and outputs of each item.
- Construct functional and reliability block diagrams. These must be collected or generated to show all system interfaces. The operation, interdependencies, and interrelationships of all system functional entities must be fully understood in order to perform adequate failure analysis.
- Identify all potential item and interface failure modes, and define their effect on the immediate function or item, on the system, and on the mission to be performed.
- Evaluate each failure mode in terms of the worst potential failure modes that can result and assign a Severity Classification Category (I through IV, identical to those used in HA). Severity for FA is defined as the consequences of a failure mode. Severity, as used in FA, considers the worst potential consequence of a failure, determined by the degree of injury, property damage, or system damage that could ultimately occur.

- Identify failure detection methods and compensating provisions for each failure mode.
- Identify corrective design or other actions required to eliminate the failure or control the risk. The rationale for acceptance of Category I and Category II failure modes, in which all reasonable actions and considerations that have been accomplished to reduce occurrence of a given failure mode, and provide a qualitative basis for acceptance of the design, will address the following:
  - Design: Those features of the design that relate to the identified failure mode that minimizes the occurrence of the failure mode, that is, safety factors, parts derating criteria, etc.
  - Test: Those tests accomplished that verify the design features and tests at hardware acceptance or during ground turnaround or maintenance that would detect the failure mode occurrence.
  - Inspection: The inspection accomplished to ensure that the hardware is being built to the design requirements and the inspection accomplished during turnaround operations or maintenance that would detect the failure mode or evidence of conditions that could cause the failure mode.
  - History: A statement of history relating to this design or a similar design.
- Identify effects of corrective actions or other system attributes, such as requirements for logistics support.
- Document the analysis and summarize the problems that could not be corrected by design, and identify the special controls that are necessary to reduce failure risk.

FMECA adds Criticality Analysis (CA) to the FMEA. The difference between these two safety techniques is best understood in terms of the following equation:

$$FMECA = FMEA + CA$$

Criticality Analysis (CA), as defined in MIL-STD-1629A, involves the calculation of the Criticality Number. This number is further differentiated into the Failure Mode Criticality Number (C_m) and the Item Criticality Number (C_r):

Failure Mode Criticality Number:

$$C_m = \beta \alpha \lambda_p t$$

where
C_m = Criticality number for failure mode
β = Conditional probability of mission loss, as follows:

Failure Effect	β Value
Actual loss	1.00
Probable loss	>0.10 to <1.00
Possible loss	>0.0 to 0.10
No effect	0

and

α = Failure mode ratio: the probability expressed as a decimal fraction that the part or item will fail in the identified mode. Failure mode ratio source data may be operational historical data or analytical data from bench tests, field tests, simulations, or analyst judgment, based on analysis of the item's functions.

λ_p = Part failure rate: the part failure rate, per MIL-STD-1629A, is obtained from "the appropriate reliability prediction." These predictions may exist in the System Design Document or in several industry-specific specifications. The specification instructs that application factors and environmental factors shall be used to illustrate differences in operating stresses. Many of these factors will be identified in the test and evaluation effort as the life cycle matures.

t = Operating time

Item Criticality Number:

$$C_r = \Sigma_{n-1}^{j}(\beta\alpha\lambda_p t)_n, \, n = 1, 2, 3, \ldots, j$$

where

C_r = Criticality number for the item

n = Failure modes of the items that fall under a particular Criticality classification

j = Last failure mode of the item in that particular Criticality classification

Figure 3.2 shows a sample FMEA worksheet, and Figure 3.3 illustrates a sample CA worksheet per MIL-STD-1629A. Criticality is defined as a relative measure of the consequences of a failure mode and its frequency of occurrences. Early in the system design, in order to perform a useful Criticality Analysis (CA), parts configuration data and component failure rate data must be collected.

The FTA safety technique is a quantitative, deductive reasoning (i.e., flows from general to specific) technique, also referred to as a "drill-down" technique, and is used to evaluate the effects of individual and multiple hardware and software faults, interfaces, environmental conditions, and human error on the system. FTA is versatile in that, early in the design phase, fault tolerance (i.e., the degree of faults that may exist before a design or operational change must be made) and fault injection (i.e., a deliberate introduction of a fault on a system or subsystem to see the result) may be investigated. In addition, FTA can be used as an accident analysis tool to validate the hypothesis of the most probable cause of an undesired event that actually occurred. Because the probabilities assigned represent actual system values, fault tree results are quantitative and provide stakeholders with a valid decision-making tool.

The *top event* of the tree is known as the undesired event. In defense applications, the top undesired event usually contains the word "Loss," that is, Loss of Aircraft, Loss of Weapon, etc. In addition, a desired event such as "Zero Accidents" may be modeled to analyze how a system must operate to obtain a desired objective. In order to build the bridge between the methodology of systems engineering and the tools of safety, we only consider undesired events.

Fault Trees are basically logic trees that show the associations between events. Figure 3.4 shows the Fault Tree concept, and Figure 3.5 shows the analysis symbology.

System ――――
Indenture Level ――――
Reference Drawing ――――
Mission ――――

Date ――――
Sheet ―― of ――
Compiled by ――――
Approved by ――――

ID NO.	ITEM/ FUNCTIONAL ID (NOMENCLATURE)	FUNCTION	FAILURE MODES AND CAUSES	MISSION PHASE/ OPERATIONAL MODE	FAILURE EFFECTS			FAILURE DETECTION METHOD	COMPENSATING PROVISIONS	SEVERITY CLASS	REMARKS
					LOCAL EFFECTS	NEXT HIGHER LEVEL	END EFFECTS				

FIGURE 3.2 Failure modes and effects analysis worksheet.

System _____
Indenture Level _____
Reference Drawing _____
Mission _____

Date _____
Sheet _____ of _____
Compiled by _____
Approved by _____

ID NO.	ITEM/ FUNC. ID	FUNCTION	FAILURE MODES AND CAUSES	MISSION PHASE OPERATIONAL MODE	SEVERITY CLASS	FAILURE PROBABILITY / FAILURE RATE DATA SOURCE	FAILURE EFFECT PROBABLITIY (β)	FAILURE MODE RATIO (α)	FAILURE RATE (λ_P)	TIME t	C_m	C_r	REMARKS

FIGURE 3.3 Criticality analysis.

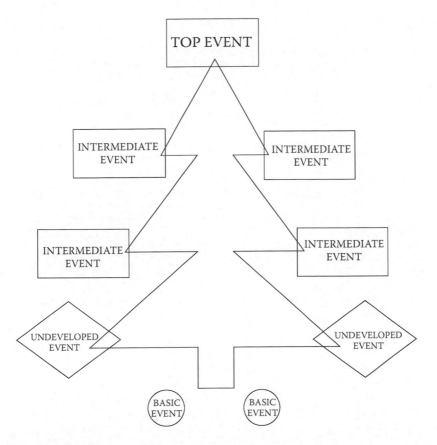

FIGURE 3.4   Fault tree concept.

Trees are constructed using logic symbols familiar to engineering and computer science undergraduates from software flowcharting or philosophy logic classes. Use of the AND gate means that all contributing events connected to the main or primary events, through the gate, must occur in order for the main or top event to occur. For example, when an AND gate is used, if only one or two of three listed events occur, then the main event will not occur. The OR gate tells the analyst that if either event connected to a main event through an OR gate occurs, the main event will also occur. This is an inclusive OR, meaning that the occurrence of any or all of the listed events will have the same result.

Figure 3.6(a) shows an example Fault Tree for an undesired top event through an OR gate. This example is taken from NASA Methodology for Conduct of Space Shuttle Hazard Program Hazard Analyses. The example shows an undesired event of an inadvertent firing of a retro-control system (RCS) jet firing. In this example, through the connection of an OR gate, the top event can occur if either

- The intermediate event of navigation software error, or
- The intermediate event of guidance and control (G&C) software error occurs, then the top event will occur.

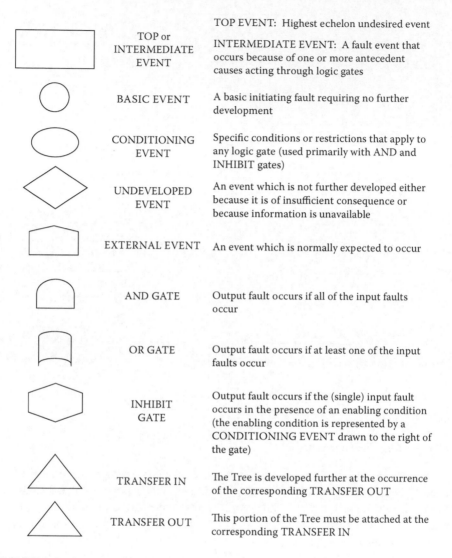

	TOP or INTERMEDIATE EVENT	TOP EVENT: Highest echelon undesired event INTERMEDIATE EVENT: A fault event that occurs because of one or more antecedent causes acting through logic gates
	BASIC EVENT	A basic initiating fault requiring no further development
	CONDITIONING EVENT	Specific conditions or restrictions that apply to any logic gate (used primarily with AND and INHIBIT gates)
	UNDEVELOPED EVENT	An event which is not further developed either because it is of insufficient consequence or because information is unavailable
	EXTERNAL EVENT	An event which is normally expected to occur
	AND GATE	Output fault occurs if all of the input faults occur
	OR GATE	Output fault occurs if at least one of the input faults occur
	INHIBIT GATE	Output fault occurs if the (single) input fault occurs in the presence of an enabling condition (the enabling condition is represented by a CONDITIONING EVENT drawn to the right of the gate)
	TRANSFER IN	The Tree is developed further at the occurrence of the corresponding TRANSFER OUT
	TRANSFER OUT	This portion of the Tree must be attached at the corresponding TRANSFER IN

FIGURE 3.5 Fault tree analysis symbols.

Also, this example shows an undeveloped event (crew error), which in this example is not developed further because it does not contribute to the occurrence of the top event.

In the example shown in Figure 3.6, if known or deduced failure rates are available, the potential effect of these rates on the top event can be numerically evaluated against the cost of controlling these risks. This example follows the fundamental concept of the "Specific Addition Rule," where the analyst simply adds the probability values for the events under an OR gate and arrives at the expected probability for the occurrence of the top event. This is seen in the following equation:

$$P(A \text{ or } B) = P(A) + P(B)$$

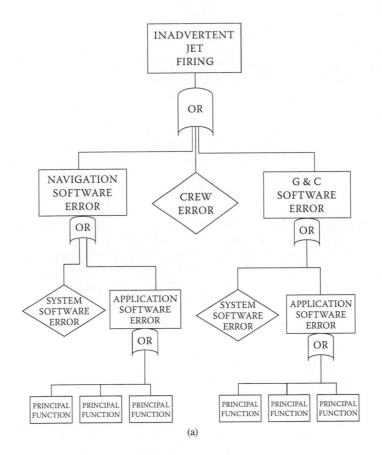

(a)

"General Addition" EXAMPLE

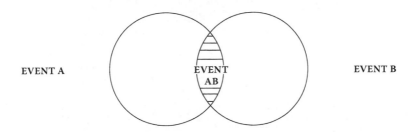

$$P(A \text{ or } B) = P(A) + P(B) - P(AB)$$

(b)

FIGURE 3.6   (a) Example fault tree—OR; (b) OR gate probability.

where

P = Probability
A = First possible event (navigation software error)
B = Second possible event (G&C software error)

The equation above is only applicable to mutually exclusive events, thus the occurrence of one has no correlation to the occurrence of the other event.

An OR gate with non-mutually exclusive events (i.e., events that overlap and may be added twice) follow the "General Addition Rule," using the following equation:

$$P(A \text{ or } B) = P(A) + P(B) - P(AB)$$

This overlapping effect is shown in Figure 3.6(b). The event labeled AB indicates that events A AND B would actually be counted twice should the events be considered mutually exclusive, and the "Specific Addition Rule" would be used instead of the "General Addition Rule."

In the case of AND gates, each listed event must occur in order to realize the top event. This is shown in Figure 3.7. To calculate the probable occurrence of an

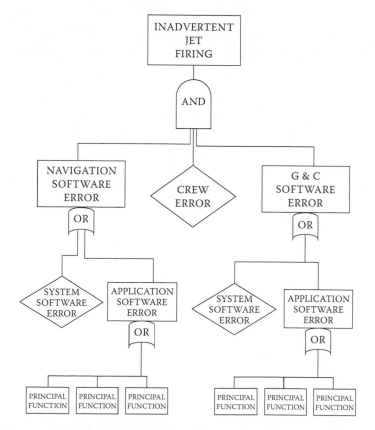

FIGURE 3.7   Example fault tree—AND.

event supported by an AND gate, the underlying event probabilities are multiplied, as shown in the following equation:

$$P(A) = P(B) \times P(C) \times P(D)$$

where

$P(A)$ = Probability of inadvertent jet firing (Top Event)
$P(B)$ = Probability of navigation software error
$P(C)$ = Probability of crew error
$P(D)$ = Probability of G & C software error

Most complex systems will need to assess probability in a more rigorous format through joint probability, conditional probability, or combinatorial mathematics. As stated, the calculation of probability for FTA is a reliability engineering function. This technique may be coordinated with the generation of the FMECA for systems with well-defined failure rate information, and performed with systems engineers, reliability engineers, and safety personnel on the team early in the life cycle.

3.2 SAFETY PROCESSES

Safety processes utilize the HA and FA techniques described above as the life cycle matures. Linearly speaking, as program requirements, design information, and test data become available, the HA is performed in the following steps.

3.2.1 PRELIMINARY HAZARD ANALYSIS (PHA)

The purpose of the PHA is to identify safety-critical areas, to identify and evaluate hazards, and to identify the safety design and operations requirements. Preparation of the PHA of a program or program item will begin at the life-cycle Concept stage (see Figure 2.1) to evaluate different concepts and develop appropriate safety requirements for early risk management decisions. Traceability between source documents and specialty engineering disciplines must be established as early in the life cycle as possible. The PHA shall identify causes down to the level at which controls are to be applied. All hazards, including those resulting from failures, regardless of subsystem or component redundancy, shall be analyzed.

As we shall see in Chapter 5, the PHA is used in support of the Preliminary Design Review (PDR) milestone to develop the requirements for new procurements while developing the statement of work or procurement specification for new hardware for the program. Completion of the PHA is required in support of the PDR to verify that the technical safety requirements have been incorporated into the preliminary design of the item for procurement. Hazards identified in the PHA will be updated in the System Hazard Analysis (SHA) and the Subsystem Hazard Analysis (SSHA) as the life cycle progresses.

The PHA provides management with knowledge of potential risks for alternative concepts during feasibility studies and program definition activities. Based on the best available data, hazards associated with the proposed design or function shall be evaluated for hazard severity, hazard probability, and operational constraint. Safety provisions and alternatives needed to eliminate hazards or reduce their associated risk to an acceptable level shall be included.

The PHA provides consideration of the following, as a minimum, for identification and evaluation of hazards:

- Hazardous components (e.g., fuels, propellants, explosives, toxic substances, hazardous construction materials, pressure systems, and other energy sources)
- Safety-related interface considerations among various elements of the system, including facilities and equipment
- Environmental constraints, including the operating environments (e.g., electrostatic discharge, vibration, noise, extreme temperatures, lightning, or fire hazards)
- Operating, test, maintenance, and emergency procedures (e.g., human error analyses of operator functions, tasks and requirements, ergonomics, equipment layout and lighting, noise, life support requirements, egress)
- Facilities, support equipment (e.g., proof-of-concept testing of hazardous systems/assemblies, which may include toxic, flammable, explosive, corrosive, or cryogenic fluids, radiation or noise emitters, and electrical power source), and training (e.g., training and certification pertaining to safety operations and maintenance)
- Safety-related equipment, safeguards, and possible alternative approaches (e.g., monitoring, interlocks, system redundancy, hardware or software fail-operational/fail-safe design considerations, subsystem protection, fire detection/suppression systems, personal protective equipment, ventilation, and noise or radiation attenuation)
- Malfunctions to the system, subsystems, or software; each malfunction shall be specified, the causing and resulting sequence of events determined, and appropriate specification and/or design changes developed

A sample PHA is shown in a columnar matrix format in Figure 3.8. This format provides a systematic method for performing the analysis, has wide applicability, and provides documented evidence of the analytical procedure. A comprehensive PHA includes

- Data Accumulation: Planned functions and interfaces must be available. Preliminary system design information, including drawings, mission descriptions, flow diagrams, operational concepts, related failure and anomaly reports, and other pertinent technical data, must be accumulated.
- Hazard Analysis: Once relevant safety data are accumulated, entries may be made in columnar matrix format. The format provides for hazard conditions, causes, effect, severity level, safety requirements, hazard elimination or control provisions, verifications, and likelihood of occurrence.

	PRELIMINARY HAZARD ANALYSIS					
PROGRAM: _____				DATE: _____		
ENGINEER: _____				PAGE: _____		
ITEM NO.	HAZARDOUS CONDITION	CAUSE	EFFECTS	PROBABILITY/ SEVERITY	ASSESSMENTS	RECOMMENDATIONS

FIGURE 3.8   Sample Preliminary Hazard Analysis (PHA).

3.2.2 SYSTEM HAZARD ANALYSIS (SHA)

The purpose of the SHA is to determine how system operation and failure modes affect the safety of the system and its subsystems. Preparation of the SHA should begin at the life-cycle Development stage (see Figure 2.1) as the system design matures, and should be updated until the design is complete. The SHA shall be generated to support the Project Milestone Critical Design Review (CDR), to be discussed in Chapter 5. Design changes will need to be evaluated to determine their effects on the safety of the system and its subsystems. This analysis should contain recommended actions, applying the system safety precedence, to eliminate or reduce the risk of identified hazards.

A sample SHA is shown in columnar matrix format in Figure 3.9. Specifically, the SHA is developed in the Development stage in conjunction with the Subsystem Hazard Analysis (SSHA) in order to examine all subsystem interfaces and interfaces with other systems for

1. Compliance with all safety criteria called out in the applicable system/ subsystem requirements documents.
2. Possible combinations of independent, dependent, and simultaneous hazard-ous events, including failures identified in the FMEA/FMECA technique, that can cause hazards to the system or personnel. Failures of controls and safety devices should be considered.
3. How normal operations of systems and subsystems can degrade the safety of the system.

ITEM NO.	HAZARDOUS CONDITION	CAUSE	EFFECTS	PROBABILITY/ SEVERITY (EXISTING)	RECOMMENDED CONTROLS	PROBABILITY/ SEVERITY (CONTROLLED)	REFERENCED DOCUMENTS

SYSTEM HAZARD ANALYSIS

System: _____

PROGRAM: _____   DATE: _____

ENGINEER: _____   PAGE: _____

FIGURE 3.9   Sample System Hazard Analysis (SHA).

4. Design changes to system, subsystems, or interfaces, logic, and software that can create new hazards to equipment and personnel.
5. Effects of reasonable human errors.
6. Determination:
   a. Of potential contribution of hardware and software (including that which is developed by external vendors/subcontractors, or commercial-off-the-shelf hardware or software) events, faults, and occurrences (such as improper timing) on safety of the system.
   b. That the safety design criteria in the hardware, software, and facilities specification(s) have been satisfied.
   c. That the method of implementation of the hardware, software, and facilities design requirements and any applied corrective actions have neither impaired nor degraded the safety of the system, nor have they introduced any new hazards.

3.2.3 SUBSYSTEM HAZARD ANALYSIS (SSHA)

The purpose of the SSHA is to verify subsystem compliance with safety requirements contained in subsystem specifications and other applicable documents; and to identify previously undiscovered hazards associated with the design of subsystems, including hardware, software, equipment, facilities, personnel actions or inactions, and hazards resulting from functional relationships between components and equipment comprising each subsystem. The SSHA is utilized to determine how operation or failure of subsystem components affects the overall safety of the system. Subsystem hazards may be caused by loss of function, accidental activation, energy source, hardware failures, software deficiencies, subsystem component interactions, inherent design flaws such as sharp edges and incompatible materials, and environmental conditions such as dust, radiation, and sand. The SSHA should identify necessary actions to determine how to eliminate or reduce the risk of identified hazards at the subsystem level.

Preparation of the SSHA should begin at the life-cycle Development stage (see Figure 2.1), when the preliminary design and concept definition are established, and further definition of the detailed design of components, equipment, and software is accomplished. The SSHA shall be updated as the result of any subsystem design change, new or modified hardware, and analyses of appropriate hardware and software failure history are generated during the life cycle.

The SSHA defines the safety-critical functions, the component fault conditions, generic hazards, and safety-critical operations and environments associated with the subsystem under the column heading "Hazardous Condition." This approach allows use of the same form for the PHA, SHA, and SSHA. Separately addressing all four hazardous conditions (generic hazards, safety-critical component fault conditions, safety-critical operations, and environment) for each SSHA provides a better opportunity to identify all hazardous conditions. A sample SSHA is shown in a columnar matrix format in Figure 3.10.

SUBSYSTEM HAZARD ANALYSIS

Subsystem: _____

PROGRAM: _____   DATE: _____

ENGINEER: _____   PAGE: _____

ITEM NO.	HAZARDOUS CONDITION	CAUSE	EFFECTS	PROBABILITY/ SEVERITY (EXISTING)	RECOMMENDED CONTROLS	PROBABILITY/ SEVERITY (CONTROLLED)	REFERENCED DOCUMENTS

FIGURE 3.10   Sample Subsystem Hazard Analysis (SSHA).

The SSHA shall include a determination for the following:

1. The modes of failure, including reasonable human errors as well as single point and common mode failures, and the effects on safety when failures occur in subsystem components.
2. The potential contribution of hardware and software (including that developed by third-party vendors) events, faults, and occurrences (such as improper timing) on the safety of the subsystem.
3. The safety design criteria in the hardware, software, and facilities specifications have been satisfied.
4. The method of implementation of hardware, software, and facilities design requirements and corrective actions have neither impaired or decreased the safety of the subsystem, nor have they introduced any new hazards or risks.
5. The implementation of safety design requirements from top-level specifications to detailed design specifications for the subsystem. The implementation of safety design requirements developed as part of the PHA shall be analyzed to ensure that it satisfies the intent of the requirements.
6. Integrated safety testing is included in hardware and software test plan and procedures.
7. System-level hazards attributed to the subsystem are analyzed, and adequate control of the potential hazard is implemented in the design.

The SSHA shall be generated in concurrence with the SHA to support the CDR, to be discussed in Chapter 5.

3.2.4 OPERATING AND SUPPORT HAZARD ANALYSIS (O&SHA)

The purpose of the O&SHA is to identify hazards and recommend risk reduction alternatives in procedurally controlled activities during all phases of intended system, hardware, and facility use. It identifies and evaluates hazards resulting from the implementation of operations or tasks performed by persons, considering the planned system configuration or state at each phase of activity; the facility interfaces; the planned environments (or ranges thereof); the supporting tools or other equipment, including software-controlled automatic test equipment specified for use; operational and task sequence, and concurrent task effects and limitations; biotechnological factors; regulatory or contractually specified personnel safety and health requirements; and the potential for unplanned events, including hazards introduced by human errors. The human shall be considered an element of the total system, receiving both inputs and initiating outputs during the conduct of this analysis.

The O&SHA must identify the safety requirements or alternatives needed to eliminate or control identified hazards, or to reduce the associated risk to a level that is acceptable under either regulatory or contractually specified criteria. It shall document the system safety assessment of procedures involved in system production, deployment, installation, assembly, test, operation, maintenance, servicing, transportation, storage, modification, and disposal.

OPERATING & SUPPORT HAZARD ANALYSIS								
System: _____							
Operational Mode:_____ Performed by: _____							
Page: _____ Date: _____							
ITEM NO.	PROCEDURE TASK	HAZARDOUS CONDITION	CAUSE	EFFECTS	PROBABILITY/ SEVERITY	ASSESSMENT	STATUS/ RECOMMENDATION

FIGURE 3.11   Sample Operating and Support Hazard Analysis (O&SHA).

Preparation of the O&SHA should begin at the life-cycle Production stage (see Figure 2.1), using all previous HAs and FAs as data for analysis of hazards that have been identified to impact system operation. A sample O&SHA is shown in columnar matrix format in Figure 3.11. The O&SHA shall identify

1. Activities that occur under hazardous conditions, their time periods, and the actions required to minimize risk during these activities/time periods
2. Changes needed in functional or design requirements for system hardware/ software, facilities, tooling, or support/test equipment to eliminate or control hazards or reduce associated risks
3. Requirements for safety devices and equipment, including personnel safety and life-support equipment
4. Warnings, cautions, and special emergency procedures (e.g., egress, rescue, render safe, etc.), including those necessitated by failure of a computer software-controlled operation to produce the expected and required safe result or indication
5. Requirements for packaging, handling, storage, transportation, maintenance, and disposal of hazardous materials
6. Requirements for safety training and personnel certification
7. Effects of nondevelopmental hardware and software across the interface with other system components or subsystems
8. Potentially hazardous system states under operator control

The O&SHA shall be generated to support the Project Milestone Final Design Review (FDR), to be discussed in Chapter 5.

Additional safety processes are performed iteratively throughout the life cycle as other specialty engineering disciplines become involved in the effort. These include Sneak Circuit Analysis (SCA) and Software Hazard Analysis (SWHA).

3.2.5 SNEAK CIRCUIT ANALYSIS (SCA)

An SCA is an analysis technique for discovering unplanned modes of operations or latent conditions that cause unexplained problems, cause unwanted functions to occur, or inhibit a desired function without regard to component failure; or cause unrepeatable or intermittent glitches or anomalies, in electrical and electronic hardware and software systems, and other systems that transfer energy such as hydraulic and pneumatic systems.

A "sneak" is a combination of conditions that causes an unexpected event. These conditions may not be detected during tests on the system. Table 3.2 shows the "Categories" and the "Potential Effects" of sneak circuits.

Figure 3.12 shows a sample sneak circuit analysis worksheet. To perform an SCA, the analysis team will require the following:

- A complete description of the system, detailing its intended purpose and design functions, such as the System Design Document
- All detailed design schematics and drawings
- Operational or Process Flow Diagrams
- Vendor certifications and specifications
- Subsystem interface descriptions

Advantages to performing an SCA are as follows:

- Costs to perform an SCA early in the design will be much less than the cost to identify and fix a sneak circuit later in the life cycle.
- The SCA will aid in the isolation of specific system faults that facilitate the performance of other analyses (i.e., FMECA).
- The SCA will reduce the time required to test a system by identifying specific problem areas ahead of time as well as those particular problems that may be created as a result of any engineering changes or system modifications.

TABLE 3.2

Sneak Circuit Analysis

Category	Potential Effect
Sneak path	May inadvertently cause current or data to flow along an unexpected route or path leading to an increase in hazard risk and a possible fault event
Sneak timing	May inadvertently cause current or data to flow at unexpected or unplanned times during system operation, which could result in system failure, damage, or loss
Sneak indication	May cause a false, inaccurate, or otherwise confusing display of system operating conditions that could result in operator error
Sneak label	Improperly labeled control sequences, operating instructions, hardware controls, etc. may lead to incorrect operator actions

SNEAK CIRCUIT ANALYSIS				
PROGRAM:_____			DATE:_____	
ENGINEER: _____			PAGE: _____	
ITEM NO.	EQUIPMENT EVALUATED	SNEAK EXPLANATION	POTENTIAL IMPACT	RECOMMENDED ACTIONS

FIGURE 3.12   Sample Sneak Circuit Analysis Worksheet.

3.2.6   SOFTWARE HAZARD ANALYSIS (SWHA)

An SWHA is an analysis technique to evaluate any potential faults in operating system and applications software requirements, codes, and programs as they may affect overall system operation.

Because complex systems require several iterations of software changes during development, regression testing, or the rerunning of test scripts to build confidence that software changes have no unintended side effects, is necessary. The Software Engineering Institute CMMI (Capability Maturity Model® Integration) provides a framework for best practices to perform fault isolation, and a model for configuration management to track changes, validated by integration testing, for system software.

3.3   SAFETY MANAGEMENT

The goal of safety management is to provide all stakeholders with a "high degree of confidence." This is defined (in Federal Railroad Administration Subpart H—Standards for Processor-Based Signal and Train Control Systems Section 236.909: Minimum Performance Standard) as follows: "as applied to the highest level of aggregation, means there exists credible safety analysis supporting the conclusion that the likelihood of the proposed condition associated with the new product being less safe than the previous condition is very small."

Safety management may be mandated by a contractual agreement between a governmental body and a vendor or subcontractor, or self-actualized within an industry depending on the stakeholder's needs. The document, shown in Figure 3.1, that guides the management of safety throughout the life cycle is the System Safety Management Plan (SSMP), known in some industries as the Product/Project Safety Plan (PSP). It may be tailored to fit the size and complexity of the stakeholder's needs. The SSMP is defined in the *US Department of Transportation Federal Transit Administration Hazard Analysis Guidelines for Transit Projects* as "a description of the planned tasks and activities to be used to implement the required system safety program. This description includes organizational responsibilities, resources, methods of accomplishment, milestones, depth of effort, and integration with other program engineering and management activities and related systems. (A document adopted by transit agencies detailing its safety policies, objectives, responsibilities, and procedures)."

The Federal Railroad Administration defines the PSP as a formal document that describes in detail all the safety aspects of the product, including but not limited to procedures for its development, installation, implementation, operation, maintenance, repair, inspection, testing, and modification, as well as analyses supporting its safety claims.

The SSMP shall define a program to satisfy the overall program system safety requirements. It shall

1. Describe the system safety organization or function within the organization of the total program using charts to show the organizational and functional relationships, roles and responsibilities, and lines of communication.
2. Describe the methods by which issues of concern are elevated directly to the program manager or the program manager's representative within the safety organization. Identify the organizational unit responsible for accomplishing each task. Identify the authority with the responsibility to resolve or accept all identified hazards.
3. Identify system safety resources throughout the duration of the life cycle to include manpower, other resources, and a summary of the qualifications of key safety personnel assigned to the effort, including those who possess coordination/approval authority for deliverables and other pertinent documentation.
4. Describe the procedures by which system safety efforts will be integrated and coordinated into overall program efforts, including the interface between system safety, systems engineering, and all other specialty engineering functions such as: quality assurance and quality control, reliability engineering, software development, maintainability, human factors engineering, and any others.
5. Describe the process through which management decisions will be made, including timely notification of unacceptable risks, necessary action, incidents or malfunctions, waivers to safety requirements, program deviations, etc.
6. Describe details of how resolution and action relative to system safety will be executed at the program management level possessing resolution authority.

In order to manage the Safety Certification Program, organizations may use a three-tier, ten-step management structure. The three-tier structure is as follows:

1. Independent Safety Assessor (ISA): The purpose of the ISA is to serve an independent role in the safety certification process, and to interface with stakeholders, vendors, subcontractors, consultants, and all other specialty engineering participants throughout the system life cycle. The ISA helps to prepare the safety certifiable items list described below, and prepares the documentation for certification of those items throughout the process. The ISA will conduct audits of the vendors, subcontractors, and consultants throughout the process, and report findings to the stakeholders accordingly.

2. System Safety Working Group (SSWG). The purpose of the SSWG is to serve as a collaborative, inter-organizational working-level assembly of subject matter experts tasked with review and approval of safety deliverables and other safety documentation including design, test, and systems integration documentation. Either in collaboration or independently from the efforts of the ISA, the SSWG shall conduct audits of the vendors, subcontractors, and consultants in order to
   a. Verify conformance of vendors and subcontractors to safety requirements, including contract deadlines and reporting processes.
   b. Ensure that regression testing of software, configuration management of design changes to hardware or software, and reporting of safety issues throughout the life cycle are properly implemented.

3. System Safety Certification Board (SSCB): Official safety certification must be presented and approved by the highest level of stakeholders within an organization. These stakeholders participate in the SSCB. The purpose of the SSCB is to
   a. Review and approve the certification process and subsequent changes.
   b. Evaluate safety evidence pertaining to certifiable items to ensure the requirements and intent of the certification process have been met.
   c. Define additional needed safety evidence.
   d. Assign responsibilities for open items resulting from SSCB actions.
   e. Hold regular meetings to assess certification process status.
   f. Issue final safety certification approval via SSCB member signatures on the Safety Certification Final Report.

Many industries follow "best practices" in the form of "10 Steps." *The US Department of Transportation Federal Transit Administration Handbook for Transit Safety & Security Certification* describes the 10 Steps to Safety Certification as follows:

Step 1: *Identify Certifiable Elements:* Elements of project deliverables: equipment, hardware, software, or documentation, are captured in the Certifiable Items List (CIL). The CIL includes all safety certifiable elements that affect safety. These elements define the scope of the project safety certification program.

Step 2: *Develop Safety Design Criteria:* As the design begins in the Concept stage (Figure 2.1), safety criteria will be identified for each certifiable item. These criteria may be traced to "lessons learned" from previous projects, existing design and performance criteria, stakeholder needs, and available standards, codes, and regulations that contribute information to help identify design elements that need to be managed through the safety process.

Step 3: *Develop and Complete Design Criteria Conformance Checklist:* As safety certifiable items are identified, a checklist is created so that compliance with safety requirements may be traced throughout the test, and validation and verification efforts. This will ensure that all safety criteria are satisfied before the design is accepted as being safe for the duration of the life cycle.

Step 4: *Perform Construction Specification Conformance:* This process is used to verify that the as-built facilities and systems incorporate safety requirements identified in the specifications and other safety documentation, including approved changes throughout the life cycle. Construction Specification Conformance is viewed as the "other half" of Design Criteria Conformance because it

- Identifies the tests and verification methods necessary to ensure that the as-built configuration contains the safety-related requirements identified in the applicable specifications and other contract documents, and
- Provides documentation that the delivered project meets those requirements.

Step 5: *Identify Additional Safety Test Requirements:* Testing must verify the functionality, performance, and safety of the system design. As the life cycle progresses from the Concept stage to the Development stage, generation of the PHA and other efforts may identify additional safety requirements. These items must be documented, tested, and verified to be safe.

Step 6: *Perform Testing and Validation in Support of the Safety Certification Program:* The verification and validation process is discussed in Section 4.7. Depending on the size and complexity of the system life cycle, testing may include Factory Acceptance and Field Acceptance for hardware, bench testing, regression testing and integration testing for software, and "shadow mode" testing to prove the safety and performance of a new system prior to removing an existing system in use. All testing must satisfy identified safety criteria before certification can be completed.

Step 7: *Manage Integrated Tests for the Safety Certification Program:* Integrated tests are any tests that require the interface of more than one element and are designed to verify the integration and compatibility between system elements. Safety criteria identified in the generation of the SSHA and any other systems integration efforts must be included and certified in this step.

Step 8: *Manage "Open Items" in the Safety Certification Program:* As safety certification proceeds throughout the system life cycle, "open items" (i.e., items identified to have a probability and severity of concern to stakeholders that could not be reduced or eliminated to an acceptable level during the design effort) must be documented and tracked as the safety certification effort continues. Open items will be tracked in the Hazard Log, and each item will have an associated Safety Mitigation Form (SMF), showing that an acceptable level of risk has been obtained through procedure, training, or another method acceptable to the stakeholders. The Risk Management process is discussed in Section 5.5.

Step 9: *Verify Operational Readiness:* Operational Readiness, known in the transportation industry as "Beneficial Use," verifies that

- Applicable operations, maintenance, and emergency rules, procedures, and plans have been developed, reviewed, and implemented.
- Facility and equipment operation and maintenance manuals have been reviewed, approved, and accepted by the stakeholders.
- A training plan for operations and maintenance personnel has been developed and successfully completed by affected personnel.

Step 10: *Conduct Final Determination of Project Readiness and Issue Safety Certification:* Essential stakeholders, assembled in the System Safety Certification Board described below, will determine project readiness and issue safety certification to document that

- All safety documentation has been approved and no outstanding items remain.
- All Category I and Category II hazard probability and severity levels have been eliminated or reduced to an acceptable level through design changes, operational procedures, or other acceptable means.
- A Final Safety Project Safety Certificate, including a signature block of all essential high-level stakeholders, is prepared and circulated for signature.

In Chapters 4 and 5, we shall see how safety management is incorporated into the systems engineering methodology through technical and project processes.

4 Technical Processes

4.1 ROLE OF TECHNICAL PROCESSES

The technical processes of the systems engineering methodology are used to

- Define the requirements for a system
- Transform the requirements into an effective product
- Permit consistent reproduction of the product where necessary
- Use the product to provide the required services
- Sustain the provision of those services
- Dispose of the product when it is retired from service

It is in the technical processes where systems engineers establish an interaction network with stakeholders and other specialty engineering professionals to enable effective development of safety, functionality, reliability, maintainability, usability, and other qualities that meet stakeholder requirements. It is the technical processes that lead to the creation of a full set of requirements that address the desired capabilities within the bounds of performance, environment, external interfaces, and design constraints.

4.2 REQUIREMENTS DEFINITION PROCESS

The purpose of the Requirements Definition Process is to define the requirements, that is, the needs, expectations, and desires of the system users and other stakeholders who provide services in a defined environment. These requirements are transformed into the System Requirements Document (SRD), which allows the generation of the System Design Document (SDD). The SRD is the highest-level document that the customer accepts to capture stakeholder needs. It will express the intended interaction that the system will have with its operational environment, and the requirements within are the reference against which each resulting operational service is validated.

4.2.1 CAPTURE SOURCE REQUIREMENTS

Requirements are captured from stakeholders, users, and subject matter experts as early as possible in the Concept stage of the life cycle (Figure 2.1). These requirements must be analyzed, quantified, and verified, and must be able to be modified. Analysis is done to guarantee clarity, consistency, and traceability to lower-level requirements and across system parameters. Quantification allows costs, schedules, and resources to be identified for requirements. Verification of requirements allows the creation of test plans that will prove out safety cases and satisfy fail-safe requirements, and provide a basis for completing and approving those tests during the life-cycle milestones.

Throughout the Concept stage, systems engineers will extract, clarify, and prioritize all of the written directives captured in the source requirements. These requirements will be expanded by a number of activities designed to break down the broad requirements statements and reveal the need for additional clarification, which will lead to either revision of the written source material or additional requirements.

A stakeholder is any entity (individual or organization) with a legitimate interest in the system. Typical stakeholders include users, operators, decision makers, parties to the agreement, regulatory bodies, developing agencies, support organizations, and society at-large. The stakeholder requirements govern the system development and are an essential factor in further defining or clarifying the scope of the development project. In an acquisition program, the requirements definition process provides the basis for the technical description of the deliverables in the acquisition contract. Source requirements also establish the basis for the Preliminary Hazard Analysis (PHA) and the System Safety Management Plan (SSMP).

4.2.2 INITIALIZE THE REQUIREMENTS DATABASE

The Requirements Database will be used to establish a database of baseline system requirements traceable to the source needs and requirements to serve as a foundation for later refinement and/or revision by subsequent activities in the systems engineering process. Although the generation of the Requirements Database will need to employ the services of database professionals familiar with building entity-relationship models and populating data dictionaries, specialty engineering professionals and other subject matter experts must be part of the database team. The database will represent the data flow and information models relevant to the system design. It may utilize relational or object-oriented tools, depending on stakeholder needs.

The Requirements Database must first be populated with the source documents that provide the basis for the total set of system requirements that will govern its design. Source documents used as inputs will include statements of user objectives, customer requirements documents, systems analysis, concept analyses, and all other relevant source requirement inputs depending on the size and complexity of the system. These source requirements should be entered in the database and shared with all database team members assigned to the requirements analysis team. The foundation of source requirements should, at a minimum, include the following:

- Project requirements
- Mission requirements
- Customer-specified constraints
- Interface and environmental requirements
- Unresolved problems or constraints identified in the capture of source requirements
- An audit trail of the resolution of problems identified
- V&V (validation and verification) methods required by the customer

- Traceability to source documentation
- An approval process to demonstrate that the database is a valid interpretation of user needs

4.2.3 ESTABLISH THE CONCEPT OF OPERATIONS

The Concept of Operations, or ConOps, is a document generated early in the system life cycle and is used to capture behavioral characteristics required of the system in the context of other systems with which it interfaces; and it captures the manner in which people will interact with the system for which the system must provide capabilities. Generation of the ConOps will allow the requirements analysis team to clearly understand operational needs. The rationale for performance requirements will be incorporated into the decision mechanism for later inclusion in the System Design Document and lower-level specifications. The ConOps will help identify safety cases that must be proven in the V&V test plans. The ConOps will

- Provide traceability between operational safety requirements and captured source requirements.
- Break down high-level source requirements into personnel requirements, maintenance requirements, etc.
- Provide the basis for computation of system capacity, behavior under/ overload, and mission-effectiveness calculations.

Creation of the ConOps will follow an iterative process early in the Concept stage. Progress will be measured by the following metrics:

1. Functional Flow Diagrams, which model sequences of interactions within and external to the system. These diagrams will help to create the PHA and the Failure Modes and Effects Criticality Analysis (FMECA)
2. Number of system external interfaces
3. Number of scenarios defined
4. Number of unresolved source requirements statements
5. Number of significant dynamic inconsistencies discovered in the source requirements

4.3 REQUIREMENTS ANALYSIS PROCESS

The purpose of the Requirements Analysis Process is to transform the stakeholder, requirement-driven view of desired services into a technical view of a product that could deliver those services. This technical view will result in the SDD and the Specification Tree, which will graphically show the deliverables associated with system requirements. Figure 4.1 illustrates the steps of the Requirements Analysis Process. It may be tailored to fit the size and complexity of the project.

System requirements are the foundation of the system definition and form the basis for the architectural design, integration, and verification. To assess the magnitude of system requirements with respect to the cost of both including, satisfying,

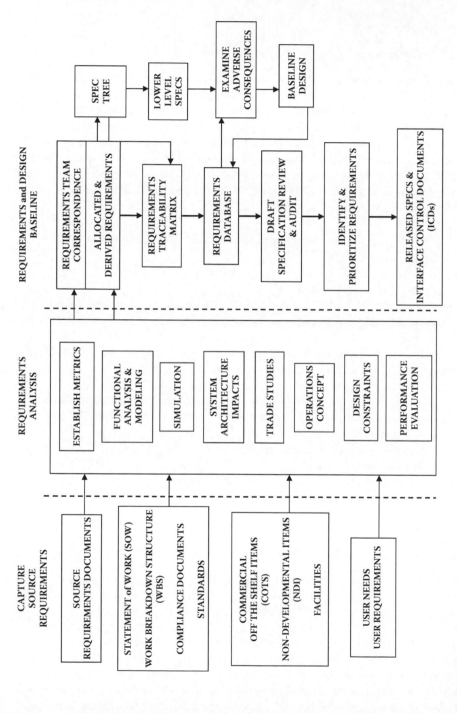

FIGURE 4.1 Requirements Analysis Process.

and/or changing the requirement, the Requirements Analysis Process is both itera-tive and recursive:

- Iterative requirements analysis is the application of the same process or set of processes that are repeated on the same level of the system. Iteration gen-erates information in the form of questions with respect to requirements, analyzed risks, or opportunities. Such questions should be resolved before completing the activities of a process or set of processes.
- Recursive requirements analysis is the application of the same process or set of processes that are applied to successive levels of system elements within the system structure. The outcomes from one application are used as inputs to the next lower (or higher) system in the system structure to arrive at a more detailed or mature set of outcomes. Such an approach adds value to successive systems in the system structure.

The output of the Requirements Analysis Process adds the verification criteria to the defined stakeholder requirements. It must be compared for traceability to and consistency with the stakeholder requirements before being used to drive the Architectural Design Process.

4.3.1   SELECTION OF REQUIREMENTS (CHARACTERISTICS OF GOOD REQUIREMENTS)

To select good requirements, they must be well written. The verb tense and mood of well-written requirements follow the verb "to be," as follows:

- "Shall": Requirements are demands upon the designer or implementer and the resulting product, and the imperative form of the verb, "shall," shall be used to identify the requirement.
- "Will": A statement containing "will" identifies a future happening. It is used to convey an item of information, explicitly not to be interpreted as a requirement. "The operator will bypass the controls by…" conveys an item of information, not a requirement on the designer of the product. Some industries have dropped the distinction between "shall" and "will" and treat either word as a means of stating a requirement. (The US Department of Defense uses *The Government Printing Office Style Manual* to write pro-curement contracts containing requirements with instructions for the use of the verb "to be." "Shall" denotes a mandatory action and "will" denotes an optional action or an action in the future.)
- "Must": "Must" is not a requirement, but is considered a strong desire by the customer, possibly to identify a system goal. "Shall" is preferable to the word "must," and only "shall" statements are verifiable and have to be veri-fied in test and performance. If both "shall" and "must" are used in a set of requirements, there is an implication of difference in degree of responsibil-ity upon the implementer.

In addition to an analysis process that assures well-written requirements, requirements must also be

- Complete: The set of requirements contains everything pertinent to the definition of system or system element being specified.
- Consistent: The set of requirements is not contradictory or duplicated and uses the same term for the same item in all requirements.
- Affordable: The set of requirements can be satisfied by a solution that is obtainable within life-cycle cost, schedule, and technical constraints.
- Bounded: The set of requirements maintains the identified scope for the intended solution without increasing beyond what is needed to satisfy user requirements.

4.3.2 DEFINE SYSTEM CAPABILITIES AND PERFORMANCE OBJECTIVES

Using the ConOps, concepts, capabilities, and constraints of the system are used to generate performance requirements and functional requirements. Constraints may include cost and cutover of system capabilities with preexisting facilities and system elements. Concepts that contribute to performance requirements include power, communications, information technology infrastructure, and environmental and personnel considerations. Performance objectives are used to establish that requirements are verifiable. Also, resource allocation may be refined as functional and performance requirements are identified.

4.3.3 DEFINE, DERIVE, AND REFINE FUNCTIONAL/PERFORMANCE REQUIREMENTS

The result of the Requirements Analysis Process should be a baseline set of complete, accurate, nonambiguous system requirements, recorded in the Requirements Database, accessible to all parties, and documented in an approved, released System Requirements Document. Defining, deriving, and refining functional/performance requirements applies to the total system over its life cycle, including its support requirements. The functions and interfaces that characterize system performance must be formally documented and flowed down to hardware and software designers. As shown in Figure 4.1, this activity is iterative, with continuous feedback loops between the Requirements Database and the baseline design, throughout the life cycle as the level of design detail matures. These feedback loops ensure that the proper inputs and feedback to hardware and software are occurring at the system and lower levels. The result of these feedback loops is a set of system and lower-level designs that are properly allocated to hardware and software and thoroughly audited to ensure that they meet requirements and are consistent with established system development practices.

4.3.4 DEVELOP SPECIFICATION TREES AND SPECIFICATIONS

The purpose of the systems engineering technical process Develop Specification Trees and Specifications is to translate identified needs into system solutions composed of specified elements of hardware and software that may be implemented

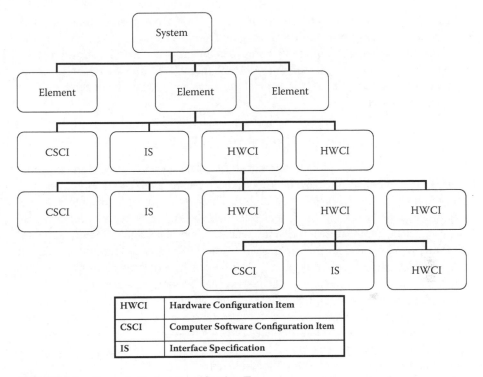

FIGURE 4.2   Example Project Specification Tree.

within the system. It involves rigorous communication and collaboration between system users, system suppliers, and the system test and evaluation group. Supporting specialty engineering disciplines monitor implementation of requirements in each area of subject matter expertise, identify requirements, and review the results of the Requirements Definition Process. The objective is to create a specification baseline for each of the configuration items at a particular level of system design (e.g., hardware, software, and operations) and place these specifications in a top-down hierarchy. The hierarchical representation of the set of specifications for the system under development is the Specification Tree. This tree establishes the framework for the integration and verification program. Figure 4.2 shows an example Project Specification Tree.

Completion of this activity occurs when

1. All specifications have been identified and located on the Specification Tree.
2. Each specification is adequate to proceed with the next stage of development or procurement.

Development of the Specification Tree will identify existing elements of the system, and those that must be designed, procured, or otherwise implemented. This must occur as early as possible in the system life cycle. At each element or node of the tree, a specification is written; and as the system development matures, a corresponding individual verification of the system design, performance, or safety will be performed.

Requirements must be crafted for each specification, from top to bottom, and must accommodate derived requirements emerging from the definition of each configuration item. A specification represents both a design entity and a verification entity.

As specifications are generated, a "rule of thumb" in the systems engineering methodology is that 50 to 250 functional/performance requirements per specification are appropriate.

4.3.5 Allocate Requirements and Establish Traceability

Depending on the industry and the complexity of the project, there exists a proliferation of requirements management tools that systems engineers may use to allocate requirements and establish traceability. These tools are used to capture source requirements, generate the SRD, and establish requirements traceability matrices (RTMs), which list the traces of requirements in a top-down and bi-directional path. As the system life cycle matures, increasing effort will be directed toward verification that the demonstrated capability of the system meets its requirements as expressed in allocated requirements captured in the system specifications. Traceability is achieved when all requirements at a particular level of the system hierarchy have been placed in the Requirements Database, and traced top to bottom as well as in a bi-directional view. Requirements must be traced to the verification program (e.g., plans, procedures, test cases, safety proofs, and reports) to provide closed-loop verification. Traceability should be maintained at all levels of documentation, as follows:

1. Allocate all system requirements to hardware, software, or manual operations, facilities, interfaces, services, or others as required.
2. Ensure that all functional and performance requirements or design constraints, either derived from or flowed down directly to a lower system architecture element, actually have been allocated to that element.
3. Ensure that traceability of requirements for source documentation is maintained throughout the life cycle until the verification program is completed and the system is accepted by the customer.
4. Ensure that the history of each requirement on the system is maintained and is retrievable.

4.3.6 Generate System Specification (System Design Document)

The SDD is a baseline set of complete, accurate, nonambiguous, verifiable system requirements, captured in the Requirements Database, documented in the SRD, and accessible to all parties. Every requirement in the SDD should be traceable to the stakeholder requirements and the SRD. Each individual requirement in the SDD should be

- Clear, unique, consistent, stand-alone (not coupled), and verifiable
- Traceable to an identified source requirement
- Neither redundant nor in conflict with any other known requirement
- Not biased by any particular implementation

It is the requirements analysis technical process of the systems engineering methodology that resolves potential conflicts and redundancies, and further decomposes requirements so that each applies only to a single system function.

4.4   ARCHITECTURAL DESIGN PROCESS

Because there is no unique solution to satisfy the stakeholder requirements, the Architectural Design Process is an iterative trade-off study, performed at a macroscopic scale, leading to a selected system architecture baseline as a final output. When alternative solutions present themselves, technical analysis and decisions are made as part of this process to identify a set of system elements. The overall objective is to create a System Architecture (defined as the selection of the types of system elements, their characteristics, and their arrangement) that meets the following criteria:

1. Satisfies the requirements (including external interfaces)
2. Implements the functional architecture
3. Is essentially close to an optimal solution within the constraints of time, budget, available knowledge and skills, and other resources
4. Is consistent with the technical maturity and acceptable risks of available elements

4.4.1   DEFINE SELECTION CRITERIA

Architectural design is strongly focused on analysis of alternatives and is part of the overall systems engineering methodology that includes Requirements Analysis (described in Section 4.3) and Decision Management (described in Section 5.4). An initial set of functions is identified to carry out the system's overall mission.

Requirements are derived to quantify how well the functions must be performed and to impose constraints on the system. An architecture option is then chosen to implement the functions and satisfy the requirements and constraints. The Architectural Design Process involves the mutual, iterative adjustment of functions, requirements, and architecture alternatives until an optimal solution is discovered.

4.4.2   CREATE SYSTEM ELEMENT ALTERNATIVES

The purpose of this activity is to identify and refine a set of architecture element options (including all hardware, software, information, procedures, interfaces, and personnel that make up the system), one level down from the top of the system hierarchy. These options constitute a set of building blocks from which system architecture options will be assembled.

This activity will provide documented assurance, with reasonable clarity for the set of options as a whole, that a basis has been established for efficient selection of the optimum architecture. The following measures are used to gauge the progress and completion of the system architecture element definition activity:

- Technical performance, schedule spans, costs, and risk estimates for each alternative
- Evidence that each alternative is consistent with the business case for the system

4.4.3 ARCHITECTURE SELECTION

The objective of this activity is to select or evolve the preferred System Architecture from the set of architecture options developed in the previous activities. The selected baseline System Architecture should be robust (i.e., allows subsequent, more detailed system definition to proceed with minimum backtracking as additional information is uncovered) and as close as possible to the theoretical optimum in meeting requirements, with risks eliminated or reduced to an acceptable level, and within available resources.

Systems engineers use tools such as modeling and simulation, and trade studies, during this activity. Modeling and simulation on large complex projects help to manage the risk of failure to meet system and performance requirements. This form of analysis is performed by specialty engineer subject matter experts who develop and validate the models, conduct the simulations, and analyze the results. Subjective trade studies are used to compare and rank alternatives. These use criteria of two types:

1. Go/no-go criteria
2. Criteria used to evaluate the relative desirability of each option on a proportional scale (systems engineers familiar with economic analysis would use a cost/benefit analysis in such a case)

This selection activity must be documented to show that the features and parameters of the optimal system architecture selection are adequate to support subsequent development of the system throughout the life cycle.

4.4.4 ARCHITECTURAL CONFIGURATION

When the System Architecture has been selected and determined to be an optimal solution, sufficient detail must be developed on system architecture elements to

1. Ensure that system architecture elements will perform as an integrated system within their intended environment
2. Enable subsequent development or design activity, as necessary, to fully define each element

Systems integration must be performed on the Systems Architecture Configuration Items so that external interfaces and hardware, software, and operational implementations at the next level of detail for all elements will seamlessly fit into their intended functions. System interfaces are captured in schematics, interface diagrams, tables, and drawings of interface data. Some industries mandate that an Interface Control Document (ICD) be delivered to ensure hardware form, fit, and function.

Definition of Architectural Configuration will involve the project process Configuration Management, to be discussed in Section 5.6. The result of this activity is a selected set of design concepts for Configuration Items (CIs, including selected technologies, configurations, design parameter values, and arrangements) to implement all the system elements as an integrated system. This includes documented definitions of all system interfaces and documented justification for the selected concepts.

4.5   IMPLEMENTATION PROCESS

The purpose of the Implementation Process is to realize a specified system element. The system element is constructed or adapted by processing the materials and/or information appropriate to the selected implementation technology and by employing appropriate technical specialties or disciplines. This process results in a system element that satisfies specified design requirements through verification, and also satisfies stakeholder requirements through validation.

4.5.1   IMPLEMENTATION STRATEGY

During the Implementation Process, systems engineers define fabrication and coding procedures, tools and equipment needed in system development, implementation tolerances, and the means and criteria for auditing configuration of resulting elements to the detailed design documentation. For mass production or the generation of replacement parts, the implementation strategy is defined/refined to achieve consistent and repeatable element production, and retained in the project decision database (see Section 5.5) for future use.

4.5.2   TRAINING PLAN

It is during the Implementation Process that training plans are created and implemented. In this activity,

- Data for training users on correct and safe procedures of operation and maintenance of the system are collected.
- Draft training documentation is prepared, and resources are allocated to perform training throughout the life cycle.
- Initial operators and maintainers are trained in a "pilot" program on the use of system elements that provide a human–system interface or require maintenance actions at that element level.

4.6   INTEGRATION PROCESS

The purpose of the Integration Process is to assemble a system that is consistent with the architectural design. This process combines system elements to form complete or partial system configurations in order to create a product specified in the system requirements. It is performed along with the V&V technical process (see Section 4.7).

4.6.1   INTEGRATION STRATEGY

In this activity, system elements (hardware/physical, software, and operating procedures) are integrated, and the end-to-end operation of the system build is demonstrated. System build is bottom-up. That is, elements at the bottom of the system hierarchy are integrated and verified first. This verifies that all boundaries between system elements have been correctly identified and described, including physical, logical, and human-system interfaces and interactions (physical, sensory, and cognitive).

4.6.2   INTEGRATION CONSTRAINTS ON DESIGN

To satisfy constraints on design, interim assembly configurations are verified to ensure the correct flow of information and data across internal and external interfaces to reduce risk and minimize errors and time spent isolating and correcting them.

4.6.3   INTEGRATION PROCEDURES

At the top level, systems integration is performed on the system and its elements, and on the system and interfacing external systems. The objective is to ensure that elements are integrated into the system and that the system is fully integrated into the larger program. At the system level, an interdisciplinary Integration Product Development Team (IPDT) is formed to address the internal interfaces among the elements of the system (i.e., System Build). At the subsystem level, Product Integration Teams (PITs) and Product Development Teams (PDTs) are formed to investigate external interfaces between the system and other systems (i.e., System Integration with External Systems).

4.6.4   SYSTEM BUILD

As stated, System Build is bottom-up. Tasks associated with the System Build are as follows:

1. Obtain the system hierarchy, which describes the relationship between system elements. This was shown for system specifications in the Specification Tree of Figure 4.2. In addition, obtain all design information and other data that define the system structure and its interfaces.
2. Determine the interfacing system elements.
3. Ascertain the functional and physical interfaces of the system and system elements. This will require the knowledge of specialty engineers who can define functions flowing bi-directionally, such as data, commands, and power. It will also require a detailed assessment of the physical interfaces, such as fluid flow, heat transfer, and mechanical and electrical equipment, loads, and footprints.
4. Organize Interface Control Drawings (ICDs) to document the interfaces and to provide a basis for negotiating the interfaces between the parties to the interfaces.
5. Work with fabrication/manufacturing groups to verify functional and physical internal interfaces, and to ensure changes are incorporated into the specifications.

6. Review test procedures and integration plans that verify the interfaces.
7. Audit design changes to ensure that interface changes are incorporated into specifications.

4.6.5 SYSTEM INTEGRATION WITH EXTERNAL SYSTEMS

This activity addresses the system integration external to the system (i.e., the integration of all the systems under development with interfacing external systems). As with the system build, IPDTs are utilized to create ICDs and place them under Configuration Management.

4.7 VERIFICATION AND VALIDATION (V&V) PROCESS

The primary purpose of the V&V process is to determine that system specifications, designs, processes, and products are compliant with requirements. A continuous feedback of V&V data helps reduce risk and resolve problems as early as possible in the life cycle. The goal is to complete the V&V of system capability to meet requirements prior to production and operation of the system.

4.7.1 VERIFICATION AND VALIDATION STRATEGY

A key outcome of the Project Planning Process (see Section 5.2) is the creation of project procedures and processes that specify the forms of V&V system assessments in appropriate project schedules and specifications. Verification criteria are generated as requirements are written, and a procedure to assess compliance is mapped out. Project plans, procedures, and standards must define the test resources of equipment, facilities, and personnel.

4.7.2 VERIFICATION AND VALIDATION CONCEPTS

V&V methods include inspection, analysis, demonstration, and validation and verification. V&V activities are determined by the perceived risks, safety, and criticality of the element under consideration. Use of a requirements management tool is essential once a design has been established and V&V begins. A unique requirements identifier can be used for traceability to the V&V plans, procedures, and reports to provide a closed-loop process from system capability, as proven through a V&V process back to the source requirement. Basic V&V activities are as follows:

1. *Inspection:* An examination of the item against applicable drawings and other documentation to satisfy compliance with stated item requirements. Hardware must have applicable tolerances to satisfy form, fit, and function with other components. Software modules must satisfy test scripts and check sums to prove that they will run in larger programs, as well as with other software modules.
2. *Analysis:* Analysis is the use of analytical data, modeling, or simulation to show theoretical compliance. Modeling and simulation such as computational

fluid dynamics software modeling or wind tunnel tests of system elements for high-speed, high-altitude airfoils of fighter jets are utilized when V&V to realistic conditions cannot be achieved or is not cost-effective, and when such means establish that the appropriate requirement, specification, or derived requirement is met by the proposed solution.

3. *Demonstration:* A qualitative exhibition of functional performance, usually accomplished with no or minimal instrumentation. Demonstration uses a set of V&V activities with system stimuli selected by the system developer. In the case of Communications Based Train Control (CBTC), the train control system must demonstrate that the command center can send speed orders to CBTC-equipped trains given that track circuits may be occupied by trains not equipped with CBTC equipment (work trains or trains with control systems disabled). Demonstrations are helpful when requirements or specifications are quantified in statistical terms such as Mean Time to Repair (MTTR) or Mean Time to Failure (MTTF).

4. *Test:* A V&V activity by which the operability, supportability, or performance capability of an item is verified when subjected to controlled conditions that are real or simulated. These verifications often use special test equipment or instrumentation to obtain very accurate quantitative data for analysis. There are four basic test categories:

   a. *Development test:* Conducted on new items to demonstrate proof of concept or feasibility.

   b. *Qualification test:* Conducted to prove that the system design meets its requirements with a predetermined margin above expected environmental or operating conditions. For example, the US Department of Defense uses a "slow cookoff test" to prove that explosives and propellants stored at elevated temperatures will not degrade in chemical composition over time.

   c. *Acceptance test:* Acceptance testing is conducted when the customer must accept ownership of system elements as delivered by the supplier. Acceptance testing may be broken down as follows:

      i. *Factory test:* Performed in concurrence with the customer or customer's representative on-site at the vendor's facility. In the transportation industry, passenger cars and work trains are completely tested by the supplier prior to shipment to the customer. In some industries, factory test is also known as either "first article test" or "factory acceptance test" for hardware, and "bench test" for software.

      ii. *Field test:* Performed at the customer's facility upon receipt of the first article delivered by the vendor. Used to prove similitude between the results of the factory test and the performance that the customer may expect on-site.

   d. *Operational:* V&V conducted to demonstrate that the item meets its specification requirements when subjected to the actual operational environment. For hardware, engines will be run for up to hundreds of hours to prove acceptance to operational performance requirements.

5. *Certification:* Written assurance that the product or article has been developed and can perform its assigned functions in accordance with industry standards and best practices. The System Safety Certification Board mentioned in Chapter 2 is tasked with the safety certification of the system.

4.8 TRANSITION AND CUTOVER PROCESS

The purpose of the Transition and Cutover Process is to establish a capability for the customer to provide existing services as new system elements are introduced through a cutover, or to rapidly transition services of a new system with stakeholder requirements satisfied in the operational environment.

4.8.1 TRANSITION AND CUTOVER STRATEGY

The Transition and Cutover Strategy is to install a system that has satisfied all V&V requirements, together with relevant enabling systems (e.g., computer operating systems, manpower support systems, training systems) as planned. The Transition and Cutover Process is used at each level in the system structure and in each life-cycle stage to complete the criteria established for exiting that stage. This includes applicable storage and handling processes, including attention to shelf life of replacement parts, hardware components such as batteries, and operating system version numbers.

Transition and Cutover Plans should be tracked and monitored to ensure that all activities satisfy the stakeholder needs, and the Transition and Cutover Report should include a roadmap to rectify any problems that arise during this process. Transition and Cutover must be complete before the Operations Process may begin.

The Transition and Cutover Strategy is as follows:

- Prepare the site of installation or the portion of the system selected for the phase of cutover per established procedures.
- Train the users in the proper use of the system and affirm users have the knowledge, skills, and abilities necessary to perform operation and maintenance activities. This includes complete generation, review, and distribution of training and maintenance manuals, as applicable.
- Receive final confirmation from the intended users that the system may be operated and maintained as per their needs. A Transition and Cutover Report shall be prepared to acknowledge that the system has been properly installed and verified, that all issues and action items have been resolved, and that all agreements pertaining to the development and delivery of a fully supportable system have been completely satisfied or adjudicated. For safety, this means that all hazards in the O&SHA have been eliminated or reduced to an acceptable level of risk.

4.9 OPERATION PROCESS

The purpose of the Operation Process is to use the system as designed in order to deliver service according to stakeholder requirements. The Operation Process is performed in the Utilization Stage/Support Stage of the system life cycle (see Figure 2.1).

4.9.1 OPERATION STRATEGY

The Operation Process sustains the service of the system by supplying personnel to operate the system, monitor operator-system performance, and monitor the performance of the system. If system performance falls outside acceptable parameters, corrective actions must be undertaken in accordance with the Systems Engineering Management Plan (SEMP). System performance is reported during this process as follows:

- *System performance reports* (e.g., statistics, usage data, and operational cost data): New systems are designed with the intention of performing above the level of performance of an existing system. This enhanced level of performance must be observed, maintained, and reported.
- *System trouble/anomaly reports:* By the time the system is in operation, hazards identified earlier in the life cycle are expected to be either eliminated or reduced to an acceptable level. The SSMP must monitor system anomalies or other reported forms of trouble and utilize safety personnel to recommend corrective action if those hazards fall out of the acceptable range.

4.10 MAINTENANCE PROCESS

The purpose of the Maintenance Process is to sustain the capability of the system to provide a service. This process monitors the system's capability to deliver services; records problems for analysis; takes corrective, adaptive, perfective, and preventive actions; and confirms restored capability. Early in the Concept stage of the life cycle, reliability and maintainability requirements were identified in the Design Process. The Maintenance Process must ensure that these requirements are accurate as designed and, if necessary, changes are properly documented. As with the Operations Process, the Maintenance Process is performed in the Utilization Stage/Support Stage of the system life cycle.

4.10.1 MAINTENANCE STRATEGY

Employing specialty engineering subject matter experts in operations, logistics, and reliability, availability, maintainability, and safety (RAMS) throughout the Utilization Stage/Support Stage of the system life cycle will allow for proper maintenance of the system to the end of its projected use. If operational procedures must be refined, data or drawings must be modified, or replacement parts must be redesigned, these experts will help in the analysis, redesign, and/or configuration management of the system.

At the end of the useful life of the system, the system may be redesigned or a disposal effort may be undertaken. The bridge to be built between the tools of system safety and the methodologies of systems engineering may be used during disposal or redesign as described above.

5 Project Processes

5.1 ROLE OF PROJECT PROCESSES

The Project Processes are used to establish, evolve, and execute the project plans; to assess actual achievement and progress against the plans; and to control execution of the project to fulfillment. Project Planning estimates the project budget and schedule against which project progress will be assessed and controlled. Individual Project Processes may be invoked at any time in the life cycle and at any level in the project hierarchy, as required by project plans or unforeseen events. Project Processes include Project Planning, Assessment and Control, Decision Management, Risk Management, Configuration Management, and Information Management. Table 5.1 contains the list of acronyms that pertain to Project Processes.

5.2 PROJECT PLANNING PROCESS

The purpose of the Project Planning Process is to produce and communicate effective and workable project plans. Project Planning establishes the direction and infrastructure necessary to assess and control the progress of a project, and identifies the details of the work and the right set of personnel, skills, and facilities with a schedule of need for resources from within and outside the organization. The Project Planning Process begins in the Concept stage of the life cycle and continues throughout the life of the system. Safety subject matter experts must be incorporated into project planning so that risk assessment may begin as early as possible in order to identify hazards and resolve project development areas that need special attention to resolve risk.

In the building of the bridge between the methodology of systems engineering and the tools of safety, it is essential to assemble the proper specialty engineers and subject matter experts in order to plan the project technical, safety, and quality management as follows:

- Prepare the Systems Engineering Management Plan (SEMP), the System Safety Management Plan (SSMP), and tailor the Quality, Configuration, Risk, and Information Management Plans to meet the needs of the project.
- Assemble the Integrated Product Development Team (IPDT). The IPDT will assist in interdisciplinary development efforts and break down organizational "stovepipes."
- Tailor the organizational safety tools (see Chapter 3), and coordinate roles and responsibilities between the SEMP and the SSMP to establish a systematic approach for identifying and resolving hazards, failures, and faults. Develop

TABLE 5.1

Project Processes Acronyms

Acronym	Definition
CMP	Configuration Management Plan
IMP	Information Management Plan
QMP	Quality Management Plan
RMP	Risk Management Plan
SEMP	Systems Engineering Management Plan
SSMP	System Safety Management Plan
WBS	Work Breakdown Structure

a planned approach to accomplish safety tasks, provide qualified people to accomplish the tasks, establish the authority for implementing the safety tasks through all levels of management, and allocate appropriate resources, both personnel and funding, to ensure that safety tasks are completed.

* Tailor the organization's Configuration Management processes and practices in accordance with the SEMP to establish a systematic approach to review, approve, and track change requests to drawings, documentation, and deliverables.

The two most important activities of the Project Planning Process are development of the SEMP and assembly of the IPDT.

5.2.1 SYSTEMS ENGINEERING MANAGEMENT PLAN

The SEMP includes the identification of required technical reviews and their completion criteria, methods for controlling changes, risk and opportunity assessment and methodology, and the identification of other technical plans and documentation to be produced for the project.

The SEMP identifies the activities to be accomplished, key events that must be satisfied at decision gates (e.g., Preliminary Design Review [PDR], Critical Design Review [CDR], and Final Design Review [FDR]) throughout the project, work packages that define the working schedule, and the assignment of required resources (i.e., personnel, equipment and facilities) that define project budget.

The SEMP is the top-level plan by which the systems engineering methodology is employed to manage the project and, as such, defines how the project will be organized, structured, and conducted as well as how the total engineering process will be controlled to provide a product that satisfies stakeholder requirements. The SEMP is supported by a project Work Breakdown Structure (WBS) that defines the project task hierarchy.

Generation of the SEMP should begin in the Concept stage. The document should be submitted for stakeholder review and approval, and utilized in the technical management of the project as early as possible. Creation of the SEMP involves

- Defining the systems engineering technical and project processes (see Chapter 4)
- The approach to functional analysis
- Trade studies for architectural design (see Section 4.4)
- Organizational roles and responsibilities

The SEMP also describes the structure of the project teams and outlines the project deliverables, decision database, specifications, and baselines. The SEMP may be tailored to fit the size and complexity of the project. It may be generated so that the template may be reused on future projects in the organization. Appendices are used to capture milestones, technical review schedules, and the roadmap for resolution of problems and identified hazards throughout the life cycle.

The SEMP should also describe the employment of specialty engineers and subject matter experts to map out the following information:

1. Organization of the systems engineering methodology with other parts of the organization:
   a. How does communication flow on an interdisciplinary level?
   b. How are issues of concern escalated to upper management?
2. Roles and responsibilities of key decision makers
3. Clear system boundaries and scope of the project
4. Project assumptions and constraints
5. Key technical objectives
6. Incorporation of safety management: SSMP
7. Validation and verification (V&V) strategy
8. Configuration Management
9. Quality Assurance/Quality Control (QA/QC planning and strategy)
10. Infrastructure support and resource management (i.e., facilities, tools, IT, personnel, etc.)
11. Reliability, Availability, Maintainability, and Safety (RAMS)
12. Integrated Logistics Support (ILS)
13. Electromagnetic Interference (EMI) abatement, radio frequency management, and electrostatic discharge (ESD)
14. Security
15. Producibility
16. Transportability

Decision gates in the form of project technical reviews must be described in the SEMP. Technical reviews are essential to ensure that the system being developed will meet requirements, and that the requirements are understood by the development team. Formal reviews are essential to determine readiness to proceed to the next stage of the system life cycle. This is essential for stakeholder acquisition professionals who must pay vendors for completion of work at specified stages of completion, and for project managers who must deliver documents, hardware, software, and other system components on a predetermined schedule. The number and frequency of these reviews and their associated decision gates must be tailored by project size and complexity. Experienced systems engineers and project managers will recognize the following technical review cycle as detailed in the *MIL-HDBK-338 Electronic Reliability Design Handbook,* Department of Defense:

- *Preliminary Design Review:* The PDR is conducted prior to the detailed design process to evaluate the progress and technical adequacy of the selected design approach, determine its compatibility with the performance requirements of the specification; and establish the existence and the physical and functional interfaces between the item and other items of equipment or facilities. Margins of safety between functional requirements and design provisions, and between operational stresses and design strengths for system elements, must be reviewed. As seen in Figure 2.1 the SSMP must be generated, and the PHA must be reviewed and approved before the PDR can begin.
- *Critical Design Review:* The CDR is conducted at the beginning of the Development stage (see Figure 2.1) when detail design is essentially complete and fabrication drawings are ready for release. It is conducted to determine that the detail design satisfies the design requirements established in the specification, and to establish the exact interface relationships between the item and other items of equipment and facilities. The System Hazard Analysis (SHA) and the Subsystem Hazard Analysis (SSHA) must be reviewed and approved before the CDR may begin.
- *Final Design Review:* The FDR is conducted near the end of the Development stage (see Figure 2.1) when the approval is granted to enter the Production stage. At this point, system hazards have been identified and either eliminated or reduced to an acceptable level, or a roadmap to track resolution of open hazards has been distributed.

Throughout the decision gate review process, the tools of safety must be integrated into the systems engineering methodology to ensure that

- An integrated SSMP has been prepared and implemented
- Hazard Analyses have been generated to identify and resolve potentially hazardous conditions, and these analyses are compatible with the associated Failure Modes and Effects Analysis/Failure Modes and Effects Criticality Analysis (FMEA/FMECA) as applicable
- Fail-safe provisions have been incorporated in the design
- Operational environments have been investigated, and the safety of personnel from equipment and facilities is ensured

5.2.2 Integrated Product Development Team

An IPDT is a process-oriented, integrated set of multidisciplinary subject matter experts organized into cross-functional teams, with appropriate resources and tasked with the responsibility and authority to define, develop, produce, and support a product, process, or service. IPDTs evolved from the need to consider all elements of the system life cycle, from conception to retirement. IPDTs, using best practices and continuous improvement, achieve significant process improvements resulting in

- Seamless interfaces within the system development teams
- Reduced engineering design time
- Fewer problems in transition from design to production
- Reduced development time and cost

IPDTs must be staffed with people who work well together and communicate. Team member's participation will vary throughout the life cycle, and different members may have primary, secondary, or minor support roles as the effort transitions from development of requirements, through the subsequent life-cycle stages. IPDTs must be empowered with responsibility for their products and services throughout the entire life cycle, concept through disposal, and the authority to succeed.

IPDTs must document their results and justify decisions they make to interfacing teams, the systems integration team, and to upper management. Basic IPDT activities include requirements definition and analysis, functional analysis, tradeoff analyses, systems integration, and V&V. Table 5.2 on IPDT high performance is from the *International Council on Systems Engineering (INCOSE) Systems Engineering Handbook.*

TABLE 5.2

Ten Techniques for High Performance in Integrated Product Development Teams (IPDTs)

Recommended Technique

1   Carefully select the staff—excellent people do excellent work.
2   Establish and maintain positive team interaction dynamics—everyone should know what is expected of the team and each individual; all should strive to meet commitments; interactions should be informal but efficient; and there is a "no blame" environment where problems are fixed and the team moves on.
3   Generate team commitment and buy-in, to the vision, objectives, tasks, and schedules.
4   Break down the job into manageable activities: those that can be accurately scheduled, assigned, and followed up on weekly.
5   Delegate and spread out routine administrative tasks among the team; this frees the leader to participate in technical activities, and gives every team member some administrative/managerial experience.
6   Create a "world-class" analysis and simulation capability—for requirements and performance to be better than the competition.
7   Schedule frequent team meetings with mandatory attendance for quick information exchanges— everyone is current; assign action items with assignee and due date.
8   Maintain a Team Leader's Notebook.
9   Anticipate and surface potential problems quickly.
10 Acknowledge and reward good work.

5.3   PROJECT ASSESSMENT AND CONTROL PROCESS

The purpose of the Project Assessment and Control Process is to determine the status of the project and direct project plan execution to ensure that the project performs according to plans and schedules, within projected budgets, to satisfy technical objectives. This process evaluates at periodic intervals (such as the decision gates discussed above) the progress and achievements against requirements plans and overall business objectives.

5.3.1   ASSESSMENT

The Project Assessment and Control Process is used to collect data to evaluate the adequacy of the project infrastructure, the availability of necessary resources, and compliance with project performance measures such as cost and "drop-dead" dates for completion of stages of the life cycle. The SEMP, WBS, schedule, and budget must be monitored with a robust feedback loop to stakeholders who are responsible for making key decisions based on the assessments.

5.3.2   CONTROL

Project Control involves both corrective and preventive actions taken to ensure that the project is performing according to plans and schedules and within project budgets. Project Control may trigger other technical processes such as the Decision Management Process. Safety tools must be well integrated into the systems engineering methodology for Project Control to be successful. Any hazards that are not able to be eliminated or reduced to an acceptable level, which may require additional funding, time, or resources to be rectified, must be communicated during the Project Control Process. Project teams need to identify critical areas and control them through monitoring, risk management, or configuration management. An effective feedback control process is an essential element to enable the improvement of project performance.

5.4   DECISION MANAGEMENT PROCESS

The purpose of the Decision Management Process is to select the most beneficial course of project action where alternatives exist. This process is employed at any time in the project life cycle where the demand for a stakeholder decision exists, whatever the nature or source of that demand, in order to reach specified, desirable, or optimized outcomes. Alternative courses of action are analyzed, and a roadmap is selected and executed. Decisions made, and their supporting rationale, are recorded for future scrutiny. Milestones and decision gates mark the most formal events of the Decision Management Process. Less formal decisions allow less formality; however, all decisions must be documented, along with the supporting rationale, to support future decision making.

Decision Management may be thought of as the intersection between social sciences such as psychology and philosophy, and technical areas of academic studies such as engineering and econometrics. Whether qualitative or quantitative, decision making involves a choice between alternatives. Best practices employ *trade studies*, which provide an objective foundation for selecting one of two or more alternative approaches to solve an engineering problem and support decisions in all stages of system development, from the Concept stage through Disposal. Trade studies may address any of a range of problems from selection of the high-level system architecture to selection of a subsystem item.

The steps of the trade study in the Decision Management Process may be tailored for the size and complexity of the project, and are described as follows:

1. *Frame the decision:* The first step in performing a trade study is to clearly articulate the decision that needs to be made. This includes forming a concise statement of the scope and context of the decision being addressed, and identifying and documenting any constraints that must be considered in the process.
2. *Determine screening and selection criteria:* The systems engineering methodology employs two distinct types of criteria: screening criteria ("must have") and selection criteria ("want to have"). Screening criteria identify those characteristics that are mandatory for any potential solution. Screening criteria are used to ensure the viability of alternatives prior to investing additional resources in the trade study. Screening criteria relate directly to stakeholder requirements and constraints.

   Selection criteria are used to discriminate between viable alternatives based on their respective performance across the entire criteria set. Selection criteria are selected from key desirable characteristics that are important to the stakeholders. These consistently include cost and risk. As we shall see in the section on the Risk Management Process (Section 5.5), risk may be differentiated into cost risk, schedule risk, programmatic risk, and performance risk if it appears that these vary separately among the alternatives.

   In addition to screening and selection criteria, decision makers must choose trade study criteria based on stakeholder needs as the life cycle matures. Backwards-compatibility of new software releases, off-the-shelf replacement parts of mother boards and other electronic hardware over time, the pipeline of suppliers for replacement parts, and maintainability are all important long-term. Trade study criteria need to be chosen with quantifiable answers to be used in decision models.
3. *Establish weighting values:* The weighting values for each criterion reflect its relative importance in the selection process. Values are assessed from 1 to 10, with 10 being the most critical criteria for selection. The weight given to a criterion must have consensus among the decision makers. To achieve objectivity, consensus should be reached before the alternative solutions have been identified. Commercial decision analysis tools are available for assistance in the weighting of criteria. It is best to trust the experience and skill of the decision-making team.
4. *Identify viable alternatives:* The next step in performing a trade study is the selection of a number of candidate alternative design solutions. When choosing a number of alternatives from two to "more than two," the task

of the decision-making team in the trade study is to avoid overlooking any viable alternative, and at the same time keep the cost of the task within reasonable parameters.

Alternatives deemed not viable should be eliminated as early as possible. The choice of alternatives may need to employ qualitative as well as quantitative analysis if agreement cannot be reached. Communication with stakeholders must be robust throughout this process.

5. *Evaluate alternatives:* Typical trade studies use measures to score alternatives from 1 to 10. These are multiplied against the weighting values previously agreed upon. As mentioned, several industry practices, scoring methods, and software are available to decision makers. Systems engineers must work closely with safety professionals and other specialty engineering subject matter experts to remain objective when scoring alternatives during the evaluation process. Rationale must be captured, and results must be documented for future scrutiny.

   The calculation of alternative score and alternative weight is known as the weighted utility value. The weighted total is the sum of the weighted utility values summed over all criteria for a given alternative. The preferred alternative nominally is the one with the best weighted total. A tabular matrix of alternatives, weights, and scores, or a decision tree, may be utilized by decision analysis experts to represent the evaluation of alternatives in a visual format.

6. *Conduct sensitivity analysis:* Once a best alternative has been chosen from the evaluation method above, stakeholders will want to know how sensitive the recommended alternative is to differently evaluated criteria, or to different estimates of the characteristics of that alternative. This allows stakeholders to consider other alternatives based on that sensitivity. Conducting a sensitivity analysis involves varying each utility and weight and re-computing the weighted total for each alternative to determine what would change if the utility values or weights were different. The significance of the change is best determined through dialogue among stakeholders and subject matter experts. If the decision is based primarily on scoring an individual criterion, then that score should be given extra care because it essentially determines the selection. The sensitivity analysis should concentrate on the criteria most affecting "winner" selection. Involvement of the stakeholders in this activity gives them confidence in the eventual choice and imparts useful insights to the whole team.

7. *Determine adverse consequences:* The determination of adverse consequences in the Decision Management Process often overlaps and complements the Risk Management Process. It is important to consider the adverse consequences that may be associated with the leading alternatives. These adverse consequences may have been reflected in the alternatives selected; however, to ensure that they are all considered, a separate step in the Decision Management Process is necessary. In cases where adverse consequences are determined to be risks, they are to be managed through the Risk Management Process (see Section 5.5).

8. *Report the results:* The results of trade studies conducted as part of the Decision Management Process must be captured in a formal report. Although the report format may be tailored by industry as well as the size and complexity of the project, all trade study reports should, as best practice, include
  - A summary description of each alternative solution
  - An explanation of the screening criteria and selection criteria used
  - An explanation of the weighting values assessed
  - A description of the scores assigned to each alternative, and the weighted utility values
  - A graphical display of scores (either tabular or decision tree)
  - A summary of the sensitivity analysis
  - A list of adverse consequences deemed to be risks, and managed through the Risk Management Process

5.5 RISK MANAGEMENT PROCESS

The purpose of the Risk Management Process is to identify, analyze, and monitor the risks of the system. The Risk Management Process is a continuous process for systematically addressing risk throughout the life cycle of a system product, process, or service. It can be applied to risks related to the acquisition, development, maintenance, or operation of a system.

Risk Management is a disciplined approach to dealing with uncertainty that is present throughout the life cycle. Risks are events that, if they occur, can influence the ability of the project team to achieve project objectives and may jeopardize the successful completion of the project. Systems engineers calculate the measurement of risk using two components:

1. The likelihood that an event will occur
2. The undesirable, negative consequence of that event if it does occur

A prioritization scheme is employed to heed risks with the greatest negative effect and the highest probability of occurrence. The objective of Risk Management is to balance the allocation of resources such that the minimum amount of resources maximizes the opportunity to eliminate or reduce risk to an acceptable level.

As seen in Figure 5.1, there are four categories of risk: technical, cost, schedule, and programmatic. Arrows show typical risk relationships, and others are certainly possible.

- *Technical risk:* A potential failure to meet requirements that can be expressed in technical terms is a source of technical risk. Technical risk exists if the system fails to achieve performance requirements, or fails to meet operability, producibility, testability, or integration requirements; or fails to meet environmental protection requirements.
- *Cost risk:* Cost risk can be projected at the total project level or at a system element level. Cost risk exists if the project must devote more resources than planned to achieve technical requirements, if the project must add

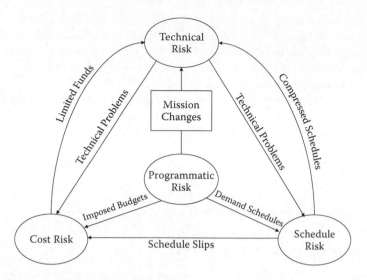

FIGURE 5.1   Risk categories and relationships.

resources to support slipped schedules due to any reason, if changes must be made to the quantity of items scheduled for production, or if organizational changes affect cost.

- *Schedule risk:* Schedule risk exists if decision gates such as PDR must slip because hazards identified in the Preliminary Hazard Analysis (PHA) cannot be reduced to an acceptable level, or if the review of safety deliverables is delayed. It may manifest itself at the system, subsystem, or item level.
- *Programmatic risk:* Programmatic risk is the result of decisions made at a higher level of management authority that may reduce the budget, scope, or level of priority of the project.

5.5.1   RISK MANAGEMENT APPROACH

Once a risk management strategy and a risk profile have been established, the three key Risk Management Process activities are Analyze Risks, Treat Risks, and Monitor Risks.

1. *Analyze risks:* Most projects combine new technology with existing systems or subsystems. Lessons learned obtained from a survey of similar upgrades or previous upgrades added to an existing system aid in analyzing risk. Analyzing risks involves identifying risks and evaluating their relative likelihood and consequence. The basis for this evaluation may be qualitative or quantitative; regardless, the objective is to set priorities and focus attention on areas of risk with the greatest consequences to the success of the project.

Safety professionals and subject matter experts must be involved in the effort to identify and analyze risks. With this participation, systems engineers define the basic characteristics of the new system as a basis for identifying past projects that are similar in technology, function, design, etc. The data collection effort to map existing risks and lessons learned to new risks must be used to further define the system for comparison purposes. This data collection effort must be documented and be defensible. Data collection may involve a review of existing failure data, and interviews with existing system experts may be conducted.

To transform qualitative expressions of probability of failure and consequences of failure into quantitative distributions or other forms of discrete, traceable forms of risk analysis information, risk must be modeled to characterize cumulative probability curves with the probability of failure and the consequences of failure expressed quantitatively in measurable terms. It is important to quantitatively characterize risk because an invalid assessment could lead to an improper conclusion with misapplied resources.

2. *Treat risks:* Risk treatment approaches need to be established for risks of greatest concern. A Risk Treatment Plan, or Risk Mitigation Action Plan, must be generated to identify the risk treatment strategy, the trigger points for action, and any other information to ensure that the treatment of risk is effectively executed. The highest technical, schedule, and cost risks must be treated as successfully as possible to ensure success of the project. There are four basic approaches to treat risk:

   a. *Avoid the risk through redesign or change of requirements.* This is also known as "requirements scrubbing," where requirements that significantly complicate the system can be scrutinized to ensure that they deliver value equivalent to their investment. Find alternative requirements that deliver the same or comparable capability.

   b. *Accept the risk "as-is."* That is, the risk as identified and analyzed is found to be at an acceptable level.

   c. *Control the risk by expending budget and other resources to reduce likelihood and/or consequence of the risk.* The trade study approach mentioned above may be used to find the alternatives that control the risk with the same or similar budget and resources.

   d. *Transfer the risk by agreement with another party that it is in their scope to mitigate the risk.* In the US Department of Defense, this has been referred to as the "pinning the rose" approach.

3. *Monitor risks:* Each risk category has certain indicators that may be used to monitor project status for treatment of risk. Technical risk may be indicated by key system technical parameters. The value added by a technical parameter versus the expenditure of resources to achieve that value throughout the life cycle over time may be tracked and evaluated. Cost and schedule risks are monitored using analysis of the cost and schedule reporting system or

another acceptable project management technique. Variances in schedule and cost must be tracked to locate opportunities to reduce the variance, or escalate the cost and schedule risk for mitigation.

5.6 CONFIGURATION MANAGEMENT PROCESS

The purpose of the Configuration Management Process is to ensure that product functional, physical, and performance characteristics are properly identified, documented, validated, and verified to establish product integrity; that changes to these product characteristics are properly identified, reviewed, approved, documented, and implemented; and that the products produced against a given set of documentation are known. Safety professionals must be included in the Configuration Management Process to ensure that changes to safety requirements are properly documented, tested, and approved. Such activities as check-in and check-out of source code, versions of system documents, and deviations created for manufactured items are part of configuration management.

As seen in Figure 5.2, changes to original requirements will occur over the system life cycle. As the design matures early in the Concept stage, a Configuration Baseline is generated. The Configuration Baseline is the collection of hardware, software, facilities, documents, procedures, and any other deliverables that serve as a reference point to maintain development and control of the system throughout its life cycle.

The Configuration Baseline is established during the review and acceptance of requirements, design, and product specifications. The Configuration Baseline may be updated at any decision gate or at any time where the requirements, design, or deliverables mature as the system life cycle advances.

The Configuration Baseline is captured in the Configuration Management Plan (CMP). This plan is designed to ensure that the proper and intended product configuration, including the hardware components and software version, is documented

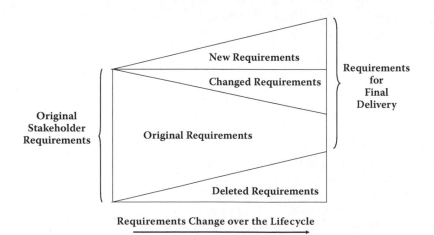

FIGURE 5.2 Requirements evolution.

and maintained throughout the life cycle of the products in use. It is in the CMP that the organizational roles and responsibilities necessary to carry out the Configuration Management Process are described. These are Configuration Identification, Configuration Control, and Configuration Audits:

- *Configuration identification:* Configuration identification is the task needed to identify system elements to be maintained under configuration control. These elements may be drawings, documents, software documentation, training manuals, and any forms of media that may need to be updated as changes to the system design affect updates to system requirements. Configuration identification uniquely identifies the elements within a baseline configuration. This unique identification promotes the ability to create and maintain master inventory lists of baselines. Identification of unique system elements allows differentiation into Configuration Items (CIs), which serve as the critical elements subjected to rigorous formal control. The compilation of the CIs is the Critical Items List. This list may reflect items to be developed, items that are to be vendor supplied, or elements of the existing system or subsystem to be integrated into the system under development. These CIs may be deliverables as negotiated in the contract, or items used to produce the deliverable items.
- *Configuration control:* Configuration control is the process by which updates to the configuration baseline are captured throughout the system life cycle. Configuration control documentation is generated, and the status of items under control is made available to the project team. Configuration control maintains integrity by facilitating approved changes and preventing the incorporation of unapproved changes into the items under configuration control. Effective configuration control requires that the extent of analysis and approval action for a proposed engineering change be in concert with the nature of the change. This is handled when an Engineering Change Proposal (ECP) is issued under Change Classification, reviewed and approved by the Configuration Control Board (CCB), and audited under the Configuration Audit Process.
  - *Change Classification:* Change Classification is the primary basis of configuration control. All changes to baseline documents are classified as either outside the scope of existing requirements or within the scope of existing requirements. A change outside the scope of project requirements is a change to a project baseline document that affects the form, fit, function, specification, reliability, or safety and will be reviewed by the submission of an ECP. A minor change that falls within the current project scope and requirements usually does not require an ECP and can be handled with the generation of an Engineering Notice (EN).
  - *Engineering Change Proposal:* A request to change the current configuration of the system is made using an ECP. An ECP may be submitted by the stakeholders, the vendors, or subcontractors throughout the life cycle of the system under development. Any circumstances that will potentially change the scope of the project or the requirements are

appropriate reasons to propose an ECP and to conduct an analysis to understand the effect of the change on existing plans, costs, and schedules. Such circumstances include

- System functionality is altered to meet a changing requirement.
- New technology or a new product extends the capabilities of the system beyond those initially documented in the stakeholder requirements.
- System life-cycle costs of development, utilization, or support are reduced.
- The reliability and availability of the system are improved.

ECPs and ENs help ensure that a system evolves in ways that allow it to continue to satisfy its operational requirements and its objectives, and that any modification is known to all interested parties.

- *Configuration Control Board:* An overall CCB is implemented at project initiation to provide an authority to coordinate, review, evaluate, and approve all proposed changes to baseline documentation and proposed changes to baseline configurations, including hardware, software, and firmware. The review board is assembled from stakeholders and specialty engineering subject matter experts. These include project managers, configuration managers, software engineers, information technology experts, systems engineers, and safety professionals.

  The chairperson of the CCB has the necessary authority to act on behalf of the project manager in all matters failing within the review board responsibility. The Configuration Manager is delegated responsibility for maintaining the status of all proposed changes. Separate review boards may be delegated the responsibility to review changes to software or hardware changes below the CI level, or those that require an EN instead of an ECP. If changes at that level require escalation, they are sent to the CCB for review and approval.

  Changes that fall within the review board jurisdiction should be evaluated for technical necessity, compliance with project requirements, compatibility with associated documents, and project impact. As changes are written while the hardware and/or software products are in various stages of development or verification, the CCB should require specific instruction so that they may identify the system impact of the proposed change. The types of impacts the CCB should assess will include the following:
  - All parts, materials, and processes are documented, reviewed, and approved for use on the project.
  - The design as documented can be produced with the methods described.
  - Project quality and reliability assurance requirements are met.
  - Interface descriptions are accurate.

- *Configuration Audit Process:* Configuration audits are performed at project milestones and decision gates to validate the control of the configuration baseline. These are independent audits performed by Configuration Management and Product Assurance teams to evaluate the evolution of a product and ensure compliance with specifications, policies, and contractual agreements. Formal audits, or Functional and Physical Configuration Audits, are performed at the completion of a product development cycle.

The Functional Configuration Audit is intended to validate that the development of a CI has been completed and has achieved the performance and functional characteristics specified in the System Design Document (SDD). The Physical Configuration Audit is a technical review of the CIs to verify as-built traceability to technical documentation. In addition, the Configuration Manager will perform in-process audits throughout the system life cycle to ensure that the Configuration Management Process is followed.

5.7   INFORMATION MANAGEMENT PROCESS

The purpose of the Information Management Process is to provide relevant, timely, complete, valid, and, if required, confidential information to designated parties during and, as appropriate, after the system life cycle. This process generates, collects, transforms, retains, retrieves, disseminates, and disposes of information. It manages designated information, including technical, project, and organizational (stakeholder and end user) information.

Information Management ensures that information is properly stored, maintained, secured, and accessible to those who need it, thereby establishing and maintaining the integrity of relevant system life-cycle information. Decision makers must be able to access information necessary to pass decision gates and document decisions made at key steps in the life cycle. The mechanism to maintain historical knowledge in prior technical and project processes of the system life cycle, such as Decision Management, Risk Management, and Configuration Management, is handled in the Information Management Process.

The initial planning efforts for Information Management are defined in the Information Management Plan. This plan establishes the scope of project information that is maintained, identifies the resources and personnel skill level required, defines the tasks to be performed, and identifies information management tools and processes, as well as methodology, standards, and procedures, that will be used on the project. Typical information includes source documents from stakeholders, contracts, project planning documents, verification documentation, engineering and safety analysis reports, and files maintained under the configuration management process.

The Information Management Process involves the following tasks:

1. Plan Information Management:
   - Identify valid sources of information. Information assets are intangible information and any tangible form of its representation, including drawings, memos, e-mail, computer files, and databases.
   - Define formats and media for capture, retention, transmission, and retrieval of information.
   - Define system storage requirements, security paradigm, access privileges, and the duration of maintenance of information storage and retrieval.

- Establish the data dictionaries and entity-relationship models of relational databases as well as the classes and structure of object-oriented databases. Information Management is most often concerned with the integration of databases, requirements management tools, and any computer-generated media such as interactive training manuals, websites, intranets, and any other forms of media that trace the history of system information throughout the project life cycle.

2. Perform Information Management:
   - Maintain information according to security and privacy requirements.
   - Retrieve and distribute information, as required.
   - Archive designated information for compliance with legal, audit, and knowledge retention requirements.
   - Retire unwanted, invalid, or unverifiable information according to organizational policy, security, and privacy requirements.

6 Management
C⁴

6.1 SKILLS, ROLES, AND RESPONSIBILITIES

When I was in graduate school over twenty years ago, my coursework at The George Washington University School of Engineering and Applied Science led to a masters degree in systems analysis and management (now known as systems engineering and management). The courses combined a study of advanced deterministic operations research skills (i.e., applied math) with courses in management science that introduced me to social sciences such as Psychology, Anthropology, Philosophy, and Sociology (i.e., in my mind at the time, "ether"), to examine how executives quantify their decision-making efforts. Terms that I learned for the first time, such as *Organizational Psychology, Decision Theory,* and *Change Management,* permeated the lectures, textbooks, and homework assignments to the point where I almost forgot that I was studying engineering at an advanced level.

To build the bridge between systems engineering and safety, and of more urgency, to avoid mistakes that have been made when the building of that bridge has not been properly performed (as we see in Chapter 7), the essential elements of management must be understood. This chapter is a brief overview of topics that could individually fill several volumes. I drew upon my years of experience as an engineering professional and the extensive research I conducted for this book to narrow down those elements to Culture, Commitment, Communication, and Coordination—or C⁴.

Management of an organization must, from the highest levels possible, instill the proper *culture, commit* to a strategy of successful achievement of stated goals, *communicate* that strategy to everyone inside and outside the organization, and *coordinate* that strategy among the stakeholders, project managers, team members, and other necessary subject matter experts in order to execute that strategy throughout the life cycle of all projects undertaken.

Just as the incorporation of these elements can contribute to a program that successfully integrates systems engineering and safety, the exclusion of any element at any point in the life cycle may result in a breakdown, where management will overlook safety, even in the face of overwhelming evidence that should have escalated the level of concern. I call this breakdown "The Glismann Effect," to be discussed in Section 6.6.

Managers must *possess skills, fill roles,* and *assume responsibilities* by having an answer to crucial questions. While no individual person can successfully perform every one of these tasks well, it is essential for complex technical organizations to assemble a management team that is endowed with a wide range of talents in as many areas as possible. It is also essential for subject matter experts in these organizations to understand and respect the contributions of other professionals with different skill sets, roles, and responsibilities.

6.1.1 Skills

Managers must develop an aptitude for competency in the following areas:

- *Technical skills:* Specialized acumen gained from years of education and experience to be recognized as a subject matter expert in one's field. In addition, this skill set needs to be mindful and respectful of the knowledge and concerns of experts in other fields as well.
- *Critical thinking skills:* Managers must develop a proficiency to make decisions; solve problems in creative, "out-of-the-box" ways; and be open to new professional experiences. They must assess situations, recognize assumptions, evaluate arguments, and draw conclusions. (For more information on developing critical thinking skills, refer to the resources on the following website: www.ThinkWatson.com).
- *Communication skills:* Communication is best defined as the transfer of information. For managers of complex technical systems, it is the link between ideas and action. The ability of managers to use all forms of communication across all areas of the project life cycle well must be developed and practiced.
- *Cooperation skills:* Managers must balance their own priorities with the concerns of others. The talent required to recognize, respect, and satisfy the concerns of others in light of the concerns of oneself and one's own team must be gained for managers to succeed in their positions.
- *Coordination skills:* Managers need to work with other people: employees, other managers, and personnel within and outside the organization. Coordination can be defined as a balanced choreography of teamwork across the various elements of an organization and throughout the life cycle.
- *Resource management skills:* Managers must understand not only the cost and trade-off of allocating resources, but also clearly understand what can and cannot be accomplished when those resources are not available. An adept understanding of cost and time allows intelligent trade-offs to be performed and communicated to the project team and supervisors throughout the life cycle.

6.1.2 Roles

Managers must comfortably embrace many tasks by playing the following parts:

1. *Interpersonal roles:* Managers must fill the roles of: Figurehead, Leader, and Liaison:
   - *Figurehead:* Manage interpersonal relationships by acknowledging requests to perform ceremonial duties that grant access to your status.
   - *Leader:* Managers are responsible for the work of the people in their area of business. Potential power that is contained in a position of

formal authority must be realized so that employees, peers, and stake-holders in other areas of the organization are motivated to follow the example of the leader.

- *Liaison:* Managers establish and maintain contacts outside the vertical chain of command. The availability of immediate and personal access to information through these contacts is priceless.

2. *Informational roles:* Managers must assume the roles of Monitor, Disseminator, and Spokesperson:

- *Monitor:* Managers are constantly scanning the environment for infor-mation, talking with liaison contacts and subordinates, and receiving unsolicited information, much of it as a result of their network of per-sonal contacts. A good portion of this information arrives in verbal form, often as gossip, hearsay, and speculation.
- *Disseminator:* Managers pass privileged information directly to subor-dinates, who might otherwise have no access to it.
- *Spokesperson:* Managers send information to people outside their orga-nizations. Although risky, this contributes to organizational visibility.

3. *Decisional roles:* Managers must take on the roles of Entrepreneur, Crisis Handler, Resource Allocator, and Negotiator:

- *Entrepreneur:* Managers seek to improve their businesses and react to opportunities. Reinvention may be necessary as events shed light on mistakes made or opportunities missed (and cases where safety con-cerns are overlooked).
- *Crisis handler:* Managers must involuntarily react to conditions.
- *Resource allocator:* Managers make decisions about who gets what, how much, when, and why.
- *Negotiator:* Managers must negotiate budget allocations, labor issues, and other formal dispute resolutions.

6.1.3 RESPONSIBILITIES

It is the responsibility of management to gain the trust and enthusiasm of the team by answering the questions of Who, What, When, Where, Why, and How, posed early on, as follows:

- Who is in charge of the project?
- What is the purpose of the project?
- What will be the project approach or methodology?
- What resources are required?
- What are the associated costs of the required resources?
- What respective contribution is expected from each participant?
- What other groups or organizations will be involved (if any)?
- What level of cooperation is expected from each group?
- When will the project be carried out?
- Where is the coordinating point for the project?
- Where is the project located?

- Why is the project needed?
- How will the project affect different groups of people within the organization?
- How will the project be tracked, monitored, evaluated, and reported?
- How do the project objectives fit the goal of the organization?

6.2 CULTURE

Culture may be described as the way we do things in organizations, including "war stories," symbols, rituals, shared values, power structures, traditions and norms, and styles. To understand corporate culture, one must recognize the type: an organization can have a power, bureaucratic, task-oriented, or person-oriented culture:

- *Power culture:* A single person or small group leads the organization. There is little or no respect for formal structures and procedures. Often these organizations are entrepreneurial. When these organizations grow, they have adaptation problems. This culture is very often difficult to change.
- *Bureaucratic culture:* All things are done following the rules. People place high value on loyalty. Political success comes from knowing how to play the system.
- *Task-oriented culture:* The organization is built around temporary project teams. It relies on people playing fairly. The problem is that situations can easily break down into vicious political infighting.
- *Person-oriented culture:* Each individual follows his or her own interests. Members have mutually beneficial links to other members.

Corporate culture can be further understood in terms of:

- *Ideas, values, and attitudes:* Culture in an organization is how we look at work, people, and situations, as well as how we respond. Managers must categorize subject matter experts in their respective fields in terms of what they think, the ideas they believe in, or the basic values they hold to be true.
- *Expected patterns of behavior:* Culture frames the groups we are expected to belong to, expected behavior, and how we are expected to feel about: "them," that is, the people who are different from us. Culture tells us what should be important as well as how to act in certain situations.

To build the bridge between systems engineering and safety in complex technical organizations, an understanding of the type of culture that exists and the effort required to enact change throughout the life cycle of a project in that culture is essential.

Managers must conduct project reviews and act on lessons learned. The commitment to safety must be pursued. Accountability at all levels of the organization must be defined to review safety documents, witness tests, and demand adherence to safety guidelines.

Changing culture is often as simple as one leader answering the question: "What can I do to help you succeed in your job, own your part, commit to achievable goals, and enjoy your work?" Enacting change to the culture to ensure the integration of safety tools into the systems engineering methodology is the legacy and responsibility of the management of the organization.

6.3 COMMITMENT

The people responsible for building the bridge between systems engineering and safety—that is, the subject matter experts in their respective fields of systems engineering, safety, and all other areas of expertise responsible for achieving project goals throughout the life cycle—must have a personal investment in the project and believe that they are empowered to act as needed. This personal investment starts with management commitment.

Commitment means that employees are confident that upper management will take an active role throughout the life cycle to define goals, secure funding and key personnel as needed, and support decisions made in order to achieve goals. Management must commit the proper authority to allow team members to move toward the desired goal, and ensure that the project's mission and objectives are validated throughout the life cycle by keeping the team on track and modifying objectives as circumstances change. When management pledges commitment, it means being active, aware, engaged, available, and willing to help the project team deliver results.

Commitment starts with proactive sponsorship that builds and sustains energy. Excellence in sponsorship plays a major role in optimizing outcomes. A sponsor initiates, funds, and supports the project from its inception through its completion and on throughout the life cycle. Managers need to spend time with all project team members, deal with misunderstandings and varying perceptions, select the right people, clarify roles, and secure the proper resources.

A common theme for the success or failure of any organizational initiative is building a guiding coalition: a bonding of sponsors and influential people who support the project.

The project sponsor is the primary risk taker, the decision maker. All stakeholders and team members know the project sponsor as the one who secures funds, assesses and manages risks, and ensures that the correct methodology is applied. It is the responsibility of the project sponsor to seek active participation, provoke discussion, and challenge achievements and deliverables. He or she establishes controls and provides content-based feedback and direction on a regular basis. It is the project sponsor who translates the personal investment of the project team to upper management, who must in turn deliver commitment to project success as well.

6.4 COMMUNICATION

As stated, communication is the transfer of information. Communication is a skill that can be learned, taught, and improved. To build the bridge between systems engineering and safety, management must understand what information must be transferred, who creates what is communicated, and what context is involved.

Management communication involves speaking, writing, listening, and employing critical thinking. Successful communication involves several different subject

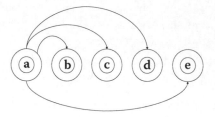

FIGURE 6.1   Communication networking.

matter experts from various parts of the organization. All groups at each stage of the project life cycle must recognize and understand the functions, goals, roles, and responsibilities within the project.

Figure 6.1 shows a network of communication across project entities. Just as the project schedule represents the sequence of activities and tasks throughout the life cycle, the elements in the sequence of the communication network are linked through effective baton-passing communication from one stage of the project to the next. Recognition and understanding at each life-cycle stage of the functions, goals, roles, and responsibilities at that stage allow stakeholders the greatest opportunity to contribute to project success.

Because communication involves one-to-many as well as many-to-many channels, networking must be employed. Networking is the process of building mutually beneficial relationships among team members. Networking should be used as a way to communicate project information bottom-up and top-down.

Management must use effective communication methods so that those who will be affected by the project directly or indirectly, as direct participants, stakeholders, or beneficiaries, should be explicitly informed as appropriate regarding the following items:

- The scope of the project
- The personnel contribution required
- The expected cost of the project, in terms of both human efforts and materials
- The merits of the project
- The project implementation plan
- The potential adverse effects of the project, if it should fail
- The alternatives, if any, for achieving the project goal
- The potential direct and indirect benefits of the project, to the organization as well as individuals

To achieve successful communication, the project manager must do the following:

- Facilitate multichannel communication interfaces
- Resolve organizational and communication hierarchies
- Encourage both formal and informal communication links

6.5 COORDINATION

To build the bridge between systems engineering and safety, successful coordination must be included along with a proper management culture that embraces safety, commitment to the incorporation of safety tools into the systems engineering methodology, and communication, where management empowers project team members by transmitting the concern for safety throughout the life cycle.

As stated, *coordination* is a balanced choreography of teamwork across the various elements of an organization and throughout the life cycle. Coordination facilitates harmonious organization of project efforts. It underscores the culture, executes the commitment, and validates the communication that management is tasked with achieving on successful projects.

Important elements of coordination include

- Balancing of tasks
- Validation of time estimates
- Authentication of lines of responsibility
- Identification of knowledge transfer points
- Standardization of work packages
- Integration of project phases
- Minimization of change orders
- Mitigation of adverse impacts of interruptions
- Avoidance of work duplication
- Identification of team interfaces
- Verification of work rates
- Validation of requirements
- Identification and implementation of process improvement opportunities

Figure 6.2 shows a coordination feedback control model. To perform coordination on the three project constraints—time, resources, and cost—Figure 6.2 shows a flowchart for project tracking and assessment of the data for feedback into project system for cost, time, and performance planning, and feed-forward into resource coordination. The loop is closed using the incoming measurement and assessment information to enhance the next phase of data collection.

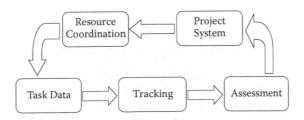

FIGURE 6.2 Coordination feedback and control model.

The four elements that management uses to achieve coordination are

1. *Responsibility Chart:* The Responsibility Chart is similar to an organizational chart in that it displays the matrix structure of an organization, specifically designed for the project life cycle, and it shows the stakeholders by organizational area and their responsibilities on the project. It is used to answer the following questions:
   - Who will do what?
   - Who will inform whom of what?
   - Whose approval is needed for what?
   - Who is responsible for which results?
   - What personnel interfaces are involved?
   - What support is needed from whom, and when?
2. *Data Analysis:* Coordination mandates that managers make decisions. Decisions require information, and information requires accurate data that must be collected, processed, and analyzed. It is when this data analysis yields valuable information that management may make decisions and empower stakeholders across functions to implement actions.
3. *Systems Integration:* Systems Integration facilitates the coordination of diverse technical and managerial efforts to enhance organizational functions, reduce costs, improve quality, improve productivity, and increase the utilization of resources. It may require adjustments to functions to permit the sharing of resources, development of new policies to accommodate product integration, or realignment of managerial responsibilities. Systems Integration includes the following important factors:
   - Physical and data interfaces between components
   - Performance metrics to be measured from the integrated system
4. *Coordination Committee:* It is through the Coordination Committee that management can improve the performance of resources allocated to the project throughout the life cycle. This committee can escalate issues as needed, reevaluate schedules, analyze performance metrics, and redesign system elements or project structure throughout the life cycle. The Coordination Committee meets to:
   - Communicate project objectives
   - Establish the agenda for project coordination
   - Define and redefine the roles and responsibilities of participants

6.6   SAFETY BREAKDOWN THEORY: THE GLISMANN EFFECT

Why did the US Navy experience an event that killed forty-seven sailors aboard a battleship? Why did NASA learn, and then presumably unlearn, lessons after the Shuttle *Challenger* tragedy in 1986, only to experience the loss of the Shuttle *Columbia* and its crew seventeen years later? Why, year after year, does any cursory search of the news result in several examples of safety concerns that are ignored or overlooked, and result in catastrophic, multi-million dollar losses, lawsuits, and lives lost?

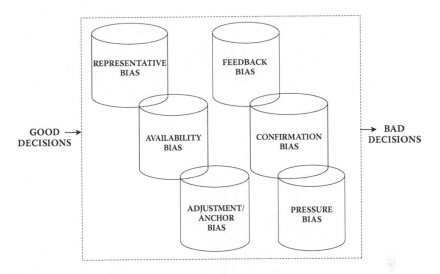

FIGURE 6.3 The Glismann Effect.

I believe that a breakdown in managerial decision making is the result of the tendency of decision makers to experience errors of judgment and choice. This occurs when they give a higher level of credibility to information that they believe is correct, regardless of the fidelity of data that supports that belief. This tendency is known as *bias*. Several biases are known to researchers and theoreticians in the Social Science fields. In the management of complex technical systems, these biases may result in an egregious discounting of safety concerns, even in the presence of strong evidence to escalate those concerns. I call this breakdown *The Glismann Effect,* as shown in Figure 6.3 and Figure 6.4. To build the bridge between systems engineering and safety, a clear understanding of this breakdown must be examined.

6.6.1 Glismann Effect: Biases

The Glismann Effect is best understood as a combination of circumstances that exists to influence the decision making of managers and employees under some state of duress. That duress may come from the situation at hand, from the influence of higher authority, or from external sources such as economic or social conditions. From a review of the existing literature on management and decision theory, this author amassed six biases that contribute to this effect. The first three biases come from the work of Daniel Kahneman, winner of the Nobel Prize in Economics, from his book *Thinking, Fast and Slow*; and the second three biases derive from various sources of literature in critical thinking and management science.

6.6.1.1 Representative Bias

Representative bias refers to the tendency by which an individual categorizes a situation based on a pattern of previous experiences or beliefs about the scenario. Representative bias can be useful when trying to make a quick decision, but it can also be limiting because it leads to close-mindedness, such as in stereotypes.

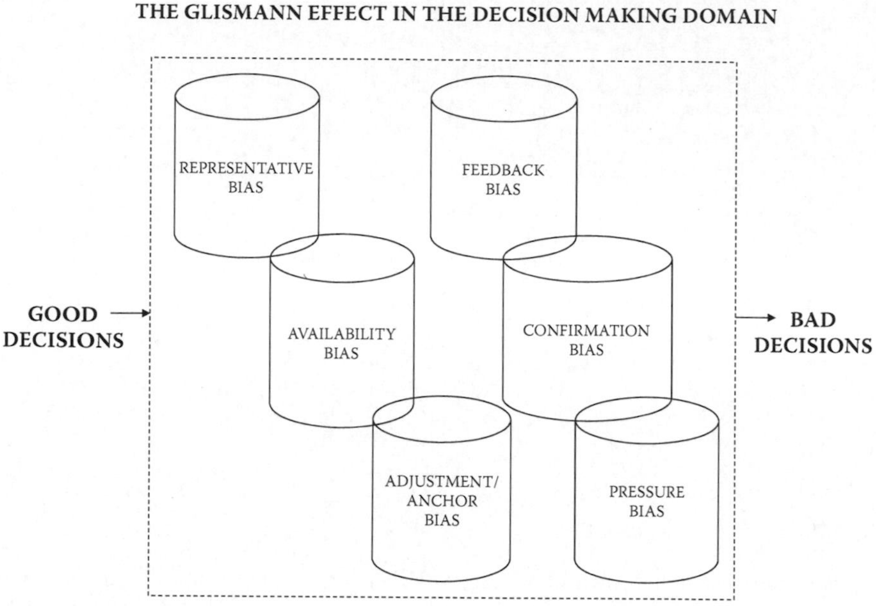

THE GLISMANN EFFECT IN THE DECISION MAKING DOMAIN

FIGURE 6.4   The Glismann Effect in the decision-making domain.

Representative bias manifests errors in judgment as follows:

1. *Excessive willingness to predict the occurrence of unlikely events:* This is an error where people will ignore statistical evidence such as "base rates" (i.e., the probability that a coin has a 50% chance of landing on "heads" or "tails" with every flip of the coin) and pay attention to the particular instance at hand. A person who buys $100 in lottery tickets believes that he is increasing his odds to win a jackpot, even though the difference between a $1 bet and a $100 bet in proportion to the number of tickets sold is negligible.
2. *Insensitivity to the quality of evidence:* A lack of discipline will make a person act on useless information as if it were any different from no information at all. A person on a vacation in Las Vegas may taxi to another casino where a friend just won $1,000 in the belief that that casino "pays out" better than the others, even though the truth is quite the opposite.

Representative bias can be alleviated with the use of Bayesian statistics, named after The Reverend Thomas Bayes, who made an early contribution to understanding the logic of how people should change their minds in the light of evidence. Bayesian statistics allows a decision-maker discipline in the decision-making process by:

- Anchoring judgment of the probability of an outcome on a plausible base rate
- Questioning the diagnosticity of evidence, that is, avoiding the sin of insensitivity with respect to the quality of evidence

6.6.1.2   Availability Bias

Availability bias is a mental shortcut in which people assess the frequency of a class or the probability of an event by the ease with which instances or occurrences can be brought to mind in order to make judgments about the probability of events. It operates on the notion that "if you can think of it, it must be important." The availability of consequences associated with an action is positively related to perceptions of the magnitude of the consequences of that action. In other words, the easier it is to recall the consequences of something, the bigger we perceive these consequences to be. Because the frequency with which events come to mind is usually not an accurate reflection of their actual probability in reality, this leads to errors in judgment.

Availability bias is a product of familiarity and salience. A General Practitioner is more likely to diagnose the next patient he sees with strep throat if several cases arrived in the office that day, due to the familiarity of the occurrence. Also, patients show up in emergency rooms reporting illnesses related to food poisoning more often than usual if news reports of food poisoning make it to the airwaves.

6.6.1.3   Adjustment/Anchor Bias

The *Adjustment/Anchor bias* describes cases in which one uses a number or a value as a starting point, known as an anchor, and adjusts said information until an acceptable value is reached. Generating a starting point, known as the "anchoring effect," occurs when people consider a particular value for an unknown quantity before estimating that quantity. It is essentially a "clutching at straws" algorithm.

Decision making under the Adjustment/Anchor bias leads to flaws because of three concerns:

1. *Insufficient adjustment:* Different starting points yield different estimates, which are biased toward the initial values. In research, the anchor is either given to a specimen, or it is based on an incomplete calculation.
2. *Evaluation of conjunctive and disjunctive events:* Conjunctive events have a linear quality, such as the completion of an undertaking (e.g., tasks in a project life cycle). For the undertaking to succeed, each of a series of events must occur. Even when each of these events is very likely, the overall probability of success can be quite low if the number of events is large. The general tendency of a decision maker to overestimate the probability of conjunctive events leads to unwarranted optimism in the evaluation of the likelihood that a plan will succeed or that a project will be completed on time.

   Disjunctive events are typically encountered in the evaluation of risks. A complex system, such as a nuclear reactor or a human body, will malfunction if any of its essential components fails. Even when the likelihood of failure in each component is slight, the probability of an overall failure can be high if many components are involved. Because of anchoring, decision makers underestimate the probabilities of failure in complex systems.

   Thus, the direction of the anchoring bias can sometimes be inferred from the structure of the event. The chain-like structure of conjunctions leads to overestimation, and the funnel-like structure of disjunctions leads to underestimation.

3. *Anchoring in the assessment of subjective probability distributions:* Subjective probability is probability derived from an individual's personal judgment about whether a specific outcome is likely to occur. Subjective probabilities contain no formal calculations and only reflect the subject's opinions and past experience.

Subjective probabilities differ from person to person. Because the probability is subjective, it contains a high degree of personal bias. An example of subjective probability could be asking New York Yankees fans, before the baseball season starts, the chances of the team winning the World Series. While there is no absolute mathematical proof behind the answer to the example, fans might still reply in actual percentage terms, such as the Yankees having a 25% chance of winning the World Series.

There are many occurrences in which a decision maker is required to express his beliefs about a known quantity (e.g., the value of the Dow-Jones Industrial Index on a particular day), in the form of a probability distribution. Such a distribution is usually obtained by asking the decision maker to select values of the quantity that correspond to specified percentiles of his subjective probability distribution. Tests have shown that decision makers do not properly calibrate their selections.

Assessing a subjective probability distribution of an uncertain quantity is a difficult task. An individual formulates his belief by adjustment from some convenient starting point. Unfortunately, the adjustment is typically not enough, and the resulting probability distribution exhibits bias due to anchoring at the starting point. The result is that the variance of estimated probability distributions turns out to be narrower than the variance of actual probability distributions. In tests, anchoring errors in the assessment of subjective probability distributions are common to both expert and non-expert respondents.

Anchoring occurs not only when the starting point is given to the subject, but also when the subject bases his estimate on the result of some incomplete calculation. In a test documented by Kahneman in which high-school students were asked to estimate, within five seconds, the calculation $1 \times 2 \times 3 \times 4 \times 5 \times 6 \times 7 \times 8$, the mean answer was 512. In a test where students were asked to estimate the calculation $8 \times 7 \times 6 \times 5 \times 4 \times 3 \times 2 \times 1$, the mean answer was 2,250. [This is known to students of mathematics as a factorial (8!), and the correct answer is: 40,320.]

This was a study of intuitive numerical estimation. To rapidly answer such questions, people may perform a few steps of computation and estimate the result by extrapolation or adjustment. Because adjustments are typically insufficient, this procedure should lead to underestimation. Furthermore, because the result of the first few steps of multiplication (performed from left to right) is higher in the descending sequence than in the ascending sequence, the former expression should be judged larger than the latter. Both predictions were confirmed.

The next three biases come from various sources of literature in critical thinking and management science.

6.6.1.4   Feedback Bias

Feedback bias is studied and defined differently by computer scientists, social scientists, and engineers. My definition is that when feedback is received in order to confirm or refine a decision made at an earlier time, the decision makers place a higher emphasis on feedback received from sources that are known to support their earlier claim more so than feedback that disputes their claim.

It is through feedback that one detects deviation from the reference point of an earlier assumption, thereby allowing for more effective system maintenance and performance. Even with accurate and timely information, there are problems that can strongly bias feedback.

In the book *Project Sponsorship* by Randall Englund and Alfonso Bucero, feedback bias may be alleviated by the use of a Feedback Action Plan and a Feedback Assessment Tool. A Feedback Action Plan allows decision makers to map out, early in the decision-making process, the sources of data to be sought, the media in which the feedback will be delivered, the timeliness of the reception of that feedback, and an action plan for use of the feedback received. A Feedback Assessment Tool is a list of feedback items that can be graded on a scale of 1 to 10 in order to allow the decision-making team to assess the quality and diversity of feedback items received, and then evaluate the changes in decisions made based on that feedback.

6.6.1.5   Confirmation Bias

Confirmation bias is a tendency to search for or interpret information in a way that confirms one's preconceptions, leading to statistical errors. Confirmation bias is a type of cognitive bias (i.e., a bias of the internal mental processes such as memory, thought, perception, and problem solving) and represents an error of inductive inference toward confirmation of the hypothesis under study.

Confirmation bias is an error in judgment wherein decision makers have been shown to actively seek out and assign more weight to evidence that confirms their hypothesis, and ignore or discount evidence that is contrary to their hypothesis. As such, it can be thought of as a form of selection bias in collecting evidence.

6.6.1.6   Pressure Bias

Pressure bias is a confluence of external influences that exerts pressure on the decision makers to discount the fidelity of information that, in the absence of that pressure, may cause them to make different decisions than those chosen. These pressures may originate from several sources:

1. *Economic sources:* In complex organizations, economic reality has influenced decisions that would not have been made in the absence of that economic pressure. Chapter 7 describes some of these pressures felt by the management of the NASA Shuttle Program, in which launches had been assigned "dollars per payload" metrics that had to be met to satisfy program costs.
2. *Political sources:* The pressure to exhibit strength as a world power during the Cold War forced the US military into a build-up that caused commanders to place soldiers in combat and deploy sailors in battle stations in

the absence of proper training or experience. The example of the explosion aboard the USS *Iowa* is examined in Chapter 7.

3. *Groupthink:* In his groundbreaking book *Groupthink: Psychological Studies of Policy Decisions and Fiascoes* (1972), Yale University Psychologist Irving L. Janis describes Groupthink as "a mode of thinking that people engage in when they are deeply involved in a cohesive in-group, when the members' strivings for unanimity override their motivation to realistically appraise alternative courses of action." Janis describes two instances from the administration of President John F. Kennedy: the "Bay of Pigs" fiasco, and, as a counterpoint, an action of many of the same group members during the successful decisions made during the Cuban Missile Crisis. With respect to decision makers elevating above pressure bias, it is important to cite Janis in that during the Cuban Missile Crisis, the President's team was successful because decision makers

   • Thoroughly canvassed a wide range of alternative courses of action
   • Surveyed the objectives and the values implicated
   • Carefully weighed the costs, drawbacks, and subtle risks of negative consequences, as well as the positive consequences, that could flow from what initially seemed the most advantageous courses of action
   • Continuously searched for relevant information to evaluate the policy alternatives
   • Conscientiously took account of the information and the expert judgments to which they were exposed, even when the information or judgments did not support the course of action they initially preferred
   • Reexamined the positive and negative consequences of all the main alternatives, including those originally considered unacceptable, before making a final choice
   • Made detailed provisions for executing the chosen course of action, with special attention given to contingency plans that might be required if various known risks were to materialize

7 Real-Life Examples

7.1 USS *IOWA* EXPLOSION

The Pacific Battleship Center, which hosts the memorial to the battleship USS *Iowa* (BB-61), can be located on a drive along South Harbor Boulevard in San Pedro, California, by nothing other than the sight of the ship itself. She floats majestically at the pier, 887 feet long, with a tall bridge (Figure 7.1) decorated with medals earned in a long service to the United States of America, and most impressively, the three gun turrets above its deck, two forward and one aft. Each turret (Figure 7.2) houses three 16-inch guns: left, right, and center. With 5-inch steel decks and 17.5-inch steel bulkheads, a turret weighs about the same as a World War II Destroyer, essentially a naval vessel all by itself.

The USS *Iowa* was originally commissioned to support World War II in 1943, de-commissioned in 1949, re-commissioned in 1951 to support the Korean War, de-commissioned again in 1958, and, to support the Reagan-era military build-up with a "600-ship Navy," the *Iowa*-class battleships (which also included the Missouri, New Jersey, and Wisconsin) were re-commissioned in 1984 at a cost of about $2 billion. No ship greater underscores the geopolitical strategy and display of military might known as "gunboat diplomacy" than a 57,450-ton battleship with nine 16-inch guns.

On the morning of April 19, 1989, while underway 330 nautical miles northeast of Puerto Rico, in participation of FLEETEX 3-89, a naval training exercise for the US Second Fleet and ships from Brazil and Venezuela, an explosion that killed forty-seven sailors occurred in the center gun of Turret II.

The US Navy produced a report, "Investigation into the 19 April 1989 Explosion in Turret II USS *Iowa* (BB-61)" dated July 15, 1989, in which it ruled out any possibility of an accidental cause and blamed the explosion on a sailor serving as the Gun Captain, in which he allegedly performed an "intentional act" of suicide and the murder of forty-six other sailors.

Hearings were then held in front of the Senate and House of Congress Armed Services Committees, after which Congress produced a report that discredited the Navy's investigation and conclusions: "USS *Iowa* Tragedy: An Investigative Failure," dated March 5, 1990.

At the request of Congress, experts from Sandia National Laboratories, one of the Department of Energy laboratories with a long reputation of expertise in the design and safety of nuclear weapons, was tasked to perform an in-depth, independent probe into possible causes of the explosion aboard the USS *Iowa*. This investigation yielded a previously unrecognized condition in which it was discovered that the explosion could have been caused accidentally, verified by field tests on the propellant utilized at the time of the tragedy. These "drop tests," designed at small-scale and full-scale

FIGURE 7.1   Bridge of the USS *Iowa*.

FIGURE 7.2   The 16-inch gun turrets I and II.

propellant load configurations and reproduced in actual land-based, 16-inch gun test facilities, yielded explosions similar to what could have happened aboard the ship.

The Navy, upon request of the Senate Armed Services Committee and after learning the results of these tests, suspended firings in the 16-inch guns. During the escalation of Desert Shield/Desert Storm, the Navy again allowed 16-inch guns to be fired with a new safety policy and more stringent training and operation controls in place. In support of Desert Storm, 16-inch guns were fired safely and without incident from the USS Missouri and USS Wisconsin. Battleships were again decommissioned soon thereafter.

The reports produced by the Navy, Congress, and Sandia as a result of this incident show a clear example of what happens when the tools of system safety are not incorporated into a systems engineering methodology. The battleships had been recommissioned without solid manning and training plans in place, and oversight of safety procedures was not exerted.

7.1.1 INCIDENT

It is best to examine what happened during this explosion by learning about the gun operations in the turret and the duties of the turret crew during a 16-inch gun firing. Figure 7.3 shows a projectile and propellant bag. Figure 7.4 represents the projectile and propellant bags as loaded into the gun breech.

The projectiles (Figure 7.3) launched from a battleship can weigh up to 2,700 pounds and hit shore-based targets over 20 miles away with pinpoint accuracy, able to penetrate 30 feet of reinforced concrete and leaving craters 50 feet wide and 20 feet deep. The normal propellant/projectile combination (Figure 7.4) is six bags of propellant, weighing about 110 pounds each, loaded behind the projectile into the gun breech by a hydraulic mechanism known as the "Rammer."

FIGURE 7.3   The 16-inch projectile.

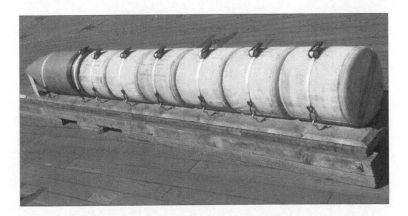

FIGURE 7.4   The 16-inch projectile/propellant breech load.

FIGURE 7.5   Cutaway of the 16-inch gun turret.

Figure 7.5 shows a cutaway diagram of a 16-inch gun turret. The gun house is the visible portion of the turret above the deck of the ship and behind the gun barrels. The gun breech, into which the Rammer loads first the projectile and then the propellant bags into the breech, is shown to the right of the guns. Below the gun house is the electrical deck, where the mechanisms that rotate the turret and control the elevation of the guns are located. Projectiles are stored and moved to the gun house below the electrical deck on the projectile decks.

The lowest area of the turret contains the powder flats. It is here that powder bags stored in hermetic canisters (as shown in Figure 7.6) are hoisted to the gun house through a powder hoist, so that propellant may be loaded onto a cradle and into the breech behind the projectile via the Rammer mechanism. Magazines partially surround the powder flat and are isolated from the turret by two heavy concentric steel bulkheads that form an annular space.

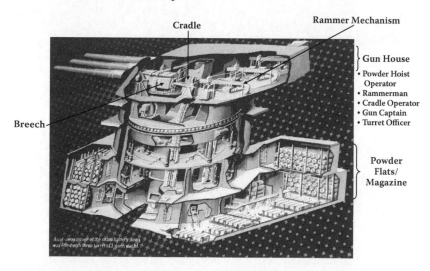

FIGURE 7.6   Propellant canister (three bags per canister).

Although the Ship Manning Document identified ninety crewmembers to be filled in a turret during General Quarters (i.e., a live-fire exercise condition), Turret II had forty-seven crewmembers at the time of the explosion, and only thirteen of them were qualified to serve in their respective roles. Key crewmembers in this incident included the

- *Turret Officer:* Stationed aft of the gun room, in which the Gun Captain, Rammerman, Cradle Operator, and Powder Hoist Operator carry out the duties of loading the gun, the Turret Officer is the supervisor of turret operations. He directs the entire crew. On April 19, 1989, the Turret Officer duty was carried out by a Chief Gunner's Mate, a 39-year-old veteran enlisted man and expert on 16-inch gun operations by the name of Reginald Ziegler.
- *Gun Captain:* Each of the nine guns has a Gun Captain. He is directly responsible for supervising gun firings and the duties of the gun crew. Because of noise during combat and firing exercises, the Gun Captain issues commands via hand signals to the Rammerman, Powder Hoist Operator, and Cradle Operator. After the explosion, it was Gunners Mate Second Class Clayton Hartwig, age 25, who was originally blamed for the "intentional act" that caused the incident. The Navy retracted this original finding in a press conference by the Chief of Naval Operations, Admiral Frank Kelso, on October 17, 1991, more than two years after the explosion.
- *Cradle Operator:* The Cradle Operator operates the hydraulic device behind the breech that positions projectiles and propellant bags to be placed on spanning trays and inserted into the gun breech by operation of the Rammer mechanism. This role was filled by Gunners Mate Third Class Richard Lawrence, age 29.
- *Powder Hoist Operator:* The Powder Hoist Operator must be experienced in operating the controls that allow propellant bags to be lifted from the powder flats below and be loaded onto the cradle so that they can be rammed into the breech behind the projectile. Second Class Boatswain Mate Gary Fisk, age 24, had never operated the powder hoist controls.
- *Rammerman:* The Rammerman, upon hand-signal commands from the Gun Captain, operates the lever that pushes, or "rams," both the projectile and the propellant into the gun breech. Controlling the distance in which the Rammer sends the propellant bags into the breech, and the speed at which the Rammer operates, are crucial to safety. An "overram," in which the propellant bags are compressed or the speed at which ramming occurs is too fast, must be avoided. Gunners Mate Third Class Robert Backherms, age 30, was in his first assignment at this post for an active firing exercise.

In the ensuing Navy investigation, it was discovered that although the 2,700-pound projectile of the center gun in Turret II was loaded successfully, some undetermined event caused a delay in the loading of the propellant bags. In addition, the Navy's investigation discovered that

1. Only thirteen watch stations (according to Navy Report Finding No. 193) in Turret II had been filled by sailors qualified to serve in those positions under the Personnel Qualification Standard (or PQS).
2. Attendance at morning quarters had been poor. It was in this meeting, known as a "pre-fire" meeting, that the crew was informed that Turret II would fire 2,700-pound projectiles with a propellant known as D846, a faster-than-normal burning propellant that is not authorized to be fired in combination with a 2,700-pound projectile.

   The D846 propellant cans for Turret II were clearly marked:

   "WARNING DO NOT USE WITH 2700 LB (AP, BLP) PROJECTILE"

3. The Captain of the ship was misinformed about the nature of the firing scheduled for center gun, Turret II. He was told that the gun would fire a "reduced charge" versus an unauthorized projectile/propellant combination.
4. The Navy Report mentioned that an overram of 21 inches occurred in the center gun of Turret II, but this overram was not a contributing factor in the explosion.

During the review of communications that occurred over the "XJ" circuit, a headphone and microphone sound system used on ships, while the load was being prepared for center gun, Turret II, Cradle Operator Lawrence stated: "I have a problem here, I'm not ready yet." In the Turret Officer's booth, Chief Ziegler is reported to have said: "Left gun loaded. Good job! Center gun is having a little trouble. We'll straighten that out." Next, Lawrence stated excitedly: "I'm not ready yet! I'm not ready yet!"

It was after this that the explosion occurred. The force of the explosion, gases produced, and extreme heat killed the forty-seven sailors instantly.

Although the remaining crew managed to heroically save the ship by fighting fires that included flooding the lower decks where the magazines were located with water for ninety minutes, much of the evidence that would have contributed to a plausible forensic analysis to determine the cause of the accident was destroyed. In addition, all witnesses in the center gun were killed, so that whatever "problem" Gunners Mate Lawrence was referring to could not be definitively determined. As water was pumped out of the turret and the crew began the recovery process, no clear record of body location was kept. Any physical evidence that was not destroyed in the explosion itself was damaged in the ensuing fire or compromised in clean-up and repair efforts after the fact.

7.1.2 ANALYSIS

Figure 7.7 shows a schematic of the D846 propellant bag. At the top is a "trim layer," a layer of additional propellant grains used to control the weight of each bag. Although in a normal load configuration, as seen in Figure 7.4, six of these bags are loaded behind the projectile, the center gun of Turret II was testing a five-bag configuration of the D846 propellant. When the breech is closed, a primer ignites the black powder

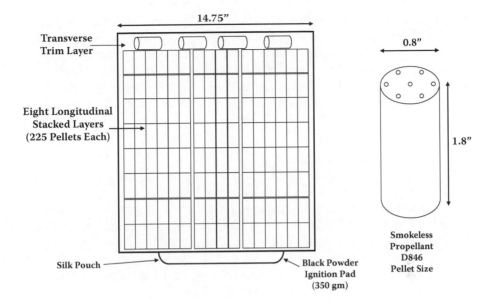

FIGURE 7.7 Schematic of D846 propellant bag.

ignition pad at the bottom of the bag closest to the breech door, and the force of ignition through the bags sends the projectile to its target.

In their independent investigation, Sandia identified that

1. An investigation of the Rammer mechanism showed that the actual overram at the time of the explosion was 4 inches greater than the Navy reported, which would have compressed the propellant bags against the projectile.
2. In testing, it was shown that the number of propellant grains (or "pellets") in the trim layer varied in number. Further testing showed that, with fewer trim layer grains than normal in an overram condition, especially at a higher-than-normal Rammer speed, these grains could begin to spark and protrude through the bag into the black powder ignition bag at the bottom of the bag above in the breech.
3. In both a "drop" test, where the projectile and propellant bags as loaded in the gun breech were suspended vertically below a weight and dropped to the ground, and in a 16-inch gun test on a ground-based Navy test facility, Sandia in cooperation with the Navy observed explosions similar to the one that occurred aboard the ship. These findings identified the reduced number of propellant grains in the trim layer, as well as the increased speed at which the overram might have occurred. Because of the condition of the physical evidence after the explosion, the actual overram speed of April 19, 1989, could not be determined.

Sandia's investigation identified a previously undiscovered safety concern: that an overram at increased speed on propellant bags with a reduced number of pellets in the trim layer (which may have been the case aboard the ship at the time of the explosion) could result in ignition of propellant grains, and inevitably cause an explosion. This caused the Navy to retract their original accusation against the Gun Captain in Turret II, and announce, in Admiral Kelso's press conference, that, "There is no clear and convincing proof of the cause of the *Iowa* explosion…despite all efforts, no certain answer regarding the cause of this terrible tragedy can be found."

7.1.3 AFTERMATH

Looking back at Chapter 6, it is clear that Navy management did not use a formal systems engineering methodology that incorporated the tools of safety into the reutilization plan when they decided to re-commission the battleships. The Navy is a bureaucratic culture with rules that adhere to the demands of orders from above, in which the expected patterns of behavior involve following those orders in the absence of personal feelings to the contrary of the consequences. In the effort to reutilize battleships in the fleet, several key concerns were overlooked in the areas of staffing, qualification, training, safety, and procedures:

Staffing:
- The existing management, manning, and training processes did not ensure that the battleships were properly manned. There were significant problems relating to the assignment of key officer and enlisted personnel to the battleships, in terms of adequate numbers, quality, and experience.

Qualification:
- Navy oversight inspections that should have assessed the *Iowa*'s PQS program failed to do so during the eighteen months preceding the explosion.
- Neither the Commanding Officer, Executive Officer, Weapons Officer, nor the Gunnery Officer aboard the USS *Iowa* knew of the large number of watch stations being manned by personnel not qualified under the PQS program.
- PQS Boards in all three turret divisions were not reviewed weekly.
- The Training Officer/PQS Coordinator did not submit monthly PQS progress reports to the Commanding Officer as required.

Training:
- At the request of Congress, the investigation by the General Accounting Office (GAO) discovered inadequate oversight inspections, the lack of a training plan for the battleship class, and significant weakness in the Navy's formal training program. The GAO testified that, "Weaknesses exist with the Navy's formal training program for 16-inch gun operations and maintenance. Limited hands-on training was being provided due to the lack of aids. Training films being used at the school were basically 1940s vintage."

- Deck department personnel assigned to turret crews were trained almost exclusively during gun firing evolution, but this training was not documented.

Safety:
- *Iowa* procedures did not ensure that safety briefs were systematically conducted for all main gun battery personnel before gunnery exercises.
- There was poor adherence to explosive safety regulations and ordnance safety.

Procedures:
- High-level officers of the Navy failed to sufficiently review and control questionable experiments with 16-inch guns, thus exposing men to possible risks that could have been minimized by using facilities designed for such experiments.
- On the day of the explosion, approximately half the people required to be present (at one of two "prefire" briefs) were absent. Musters were not taken. The Commanding Officer did not attend either one of the prefire briefs.

In addition, bias played a role in the "systemic management deficiencies" described in the ensuing investigative reports, hearings, and press conferences. These deficiencies contributed to the safety breakdown that caused this event. Specifically, feedback bias and pressure bias were clearly present.

1. *Feedback bias:* Previous successful firings of the 16-inch guns aboard the USS *Iowa*, and the discovery that the Commanding Officer did not receive monthly PQS progress reports as required, may have led him to believe that training and qualification aboard the ship was better than it actually was. Also, because the Commanding Officer was informed that a "reduced charge" test in the center gun of Turret II would be fired, rather than an "unauthorized" projectile/propellant configuration, he may not have detected a deviation from the reference point of an earlier assumption, which was that the firing would be conducted safely.
2. *Pressure bias:* Although not a contributing factor to the possible cause of the explosion that was identified in the independent investigation conducted by Sandia, a visiting Admiral was on board the USS *Iowa* at the time of the explosion. It is possible that the reason that the Commanding Officer did not attend a prefire brief on the morning of the explosion and did not know that a large number of watch stations were being manned by personnel not qualified under the PQS program may have been that his attention was focused on his esteemed visitor. This pressure bias had two major contributing factors:
   a. *Political pressure:* Navy Captains get promoted to the rank of Admiral by superior officers. As mentioned, in a bureaucratic culture with expected patterns of behavior, staffing, training, and safety problems did not contribute to upward mobility when the expectation was to witness a successful test of military superiority.

b. *Groupthink:* In the Cold War rush to build a Navy of 600 ships, having those battleships, those antiquated vessels with outdated training methods and difficulty in placing experienced personnel onboard, firing mighty weapons in a training exercise may have been more important than whether the crew was properly trained and the test was conducted properly.

The explosion aboard the USS *Iowa* represents a real-world example in which an organization, the US Navy, an organization quite familiar with the methodology of systems engineering, a well-defined management culture, and the tools of system safety, failed to build the bridge, resulting in the loss of forty-seven lives.

7.2   NASA SHUTTLE *CHALLENGER* TRAGEDY

On January 28, 1986, the Space Shuttle *Challenger*, Flight STS-51-L, which came to be known as the "Teacher in Space" mission, erupted into a ball of flames seventy-three seconds after launch. All seven astronauts on board perished after a descent of the crew compartment that lasted two and a half minutes and ended with an impact at sea at a speed of 200 miles per hour.

7.2.1   INCIDENT

In mid-April 1986, salvage efforts located several chunks of charred metal that confirmed that the tragedy was the result of an explosion of the external fuel tank. This explosion was caused by leaking gases from a faulty seal between two segments of the booster known as solid rocket motors (SRMs). This faulty seal, known as an O-ring, had allowed a stream of flame produced by the burning of hot propellant gases from the SRMs to "blow-by" or escape and ignite the external fuel tank.

A Presidential Commission (also known as the "Rogers Commission" after its head, former Attorney General William P. Rogers) was assembled to investigate the disaster.

The Commission questioned employees of the Marshall Space Flight Center in Huntsville, Alabama, who were responsible for the shuttle's entire propulsion system, including the orbiter main engine, external fuel tank, and SRBs (solid rocket boosters), and also Morton Thiokol, the contractor that had the contract on the SRMs. After the O-ring problems were discovered through internal documents leaked to the press, the Commission's stance toward NASA shifted from one of cooperative oversight to one of confrontation.

After interviews and analysis of data, the Presidential Commission released a report that found the following:

1. The *Challenger* explosion was caused by a single, catastrophic malfunction: the O-rings that were supposed to prevent hot gases from leaking out of the rocket boosters did not do their job.
2. NASA management had committed safety violations due to production pressures and launch schedules.

The Presidential Commission revealed that SRB O-rings had been a well-documented problem at NASA. The earliest documentation dated back to 1977, nearly four years before the first shuttle flight in 1981. The report was critical of NASA safety procedures. It claimed that the NASA safety system failed at monitoring shuttle operations to such an extent that the report referred to it as the "Silent Safety Program." The structure of regulatory relations within the NASA hierarchy limited the flow of information and knowledge about the O-ring problems. It inhibited the ability of safety regulators to alter the scientific paradigm on which the belief in "acceptable risk" was based.

In addition, Congress conducted hearings under the US House of Representatives Committee on Science and Technology. The Congressional Report, published in October 1986, placed the blame on people making poor technical decisions about the O-rings over a period of years: top NASA and contractor personnel who "failed to act decisively" to solve the increasingly serious anomalies in the SRB joints.

7.2.2 ANALYSIS

The reports of both the Presidential Commission and the House Committee implicated the NASA organization and Morton Thiokol in violations of safety procedures and poor decision making. Moreover, both reports were critical of the economic and political environments that were the structural origins of disaster:

- In 1982, the shuttle was deemed an "operational vehicle," no longer requiring the resources or funding needed to ensure the same level of safety as is allocated for a vehicle in development. Economics played a factor in the conditions of NASA management at the time of the launch.
- The launch had been scheduled to coincide with then-President Ronald Reagan's State of the Union speech on the night the shuttle was to be launched on January 28. Because the shuttle included Christa McAuliffe, a teacher from New Hampshire, it was believed that the President's speech was going to mention the fact that the shuttle was now a "payload-for-hire" vehicle able to include someone capable of teaching American students a lesson from space. This would have been a major boost for future funding of the space programs, which were always at the mercy of budget cuts at the federal level. Although the Presidential Commission found that no direct political pressure was enforced on NASA, the perception of being "under the gun" could not have been ignored.

Ten years after the tragedy, a Boston College sociologist, Dr. Diane Vaughan, released the book *The Challenger Launch Decision: Risky Technology, Culture and Deviance at NASA*. This book presented a revisionist history in that, from the lens and analysis of social science, it claims that, "No fundamental decision was made at NASA to do evil. Rather, a series of seemingly harmless decisions were made that incrementally moved the space agency towards a catastrophic outcome." These incremental decisions were a part of a closed culture at NASA that "normalized deviance" so that to the outside world decisions that were obviously questionable were seen by NASA's management as prudent and reasonable. In the book, Vaughan sought to answer two important questions:

1. Why, in the years preceding the *Challenger* launch, did NASA continue launching with a design known to be flawed?
2. Why did NASA launch the *Challenger* against the eve-of-launch objections of Thiokol engineers?

To gain insight into the collapse of the bridge that existed between systems engineering and safety, we first take a look at the actual technical culprit of the failure. Then, an answer to the questions above will be presented by an examination of

- The NASA culture of "acceptable risk"
- The decision-making process of the SRB Working Group
- The construction of risk
- The two internal and one external safety review teams that NASA utilized
- Flight Readiness Review (FRR)
- Events of the night before the launch

To understand the technical root cause, Figure 7.8 shows the shuttle system, including the two SRBs, external fuel tank, and shuttle main orbiter vehicle. The SRBs contribute 80% of thrust at liftoff. They burn for two minutes when, with fuel exhausted, they separate from the orbiter 24 miles downrange and drop by parachute to the sea. The SRBs are made up of a nose forward section, a shroud aft section that covers the nozzle, and a solid-propellant rocket motor (SRM). The SRM consists of four sections, each of which were cast with propellant in the Morton Thiokol factory in Utah and shipped to the Kennedy Space Flight Center in Florida for assembly.

Figure 7.9 shows a field joint cross section of a "tang and clevis" assembly. "Factory joints" are those joints between segments joined at the Thiokol factory;

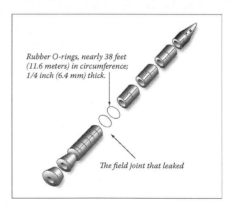

Rubber O-rings, nearly 38 feet (11.6 meters) in circumference; 1/4 inch (6.4 mm) thick.

The field joint that leaked

Shuttle System   **O-Ring Failure**

FIGURE 7.8 Shuttle system and O-ring failure.

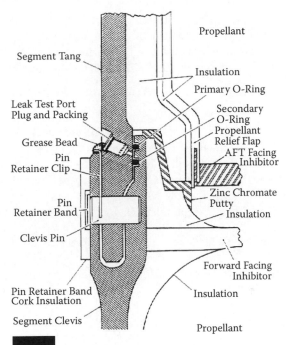

Solid Rocket Motor cross section shows positions of tang, clevis and O-rings Putty lines the joint on the side toward the propellant.

FIGURE 7.9   SRM field joint O-rings.

"field joints" are the ones between the four casting segments that are joined at Kennedy (i.e., in the field).

The tang of the top SRM fits into the deep pocket in the clevis of the lower segment. The two O-rings, which encircle the booster joints like giant rubber washers, are placed in grooves cut into the clevis inner edge before the booster segments are stacked. Made of black, rubber-like Viton®, they are about ¼ inch in diameter. When SRM segments are stacked, the tang compresses the O-rings in their grooves, as shown, so that the tiny gap between tang and clevis is closed by the O-rings. The O-rings are also known as seals because they seal the field joint against the hot gases created at ignition when the propellant inside the booster burns to provide the necessary thrust to lift the orbiter. The O-rings are protected from the hot propellant gases by asbestos-filled, zinc chromate putty, which lines the other tiny channel created when the segments are stacked.

In a number of previous shuttle flights leading up to the launch, as discovered in NASA documents by the Presidential Commission, it was reported that the rubber seal had been dangerously eroded. This condition suggested that hot gases had almost escaped. Also, O-rings were believed to be less effective in cold weather, when shrinking and cracking of the rubber would harden and not give as tight a seal, and could fail to block the escape of hot gases that could ignite the shuttle's huge external fuel tank.

The post-tragedy investigation determined that because of the unprecedented cold temperatures leading up to the launch, O-ring resiliency was impaired. Upon ignition, hot propellant gases impinged on the O-rings, creating a flame that penetrated first the aft field joint of the right SRB, and then the external tank containing liquid hydrogen and oxygen.

7.2.2.1 Acceptable Risk Process

The Acceptable Risk Process was the basis for all technical decision making at NASA, from daily decision making to the formalized, final decision process immediately prior to a launch known as the FRR. The Acceptable Risk Process was first set forth in the "Space Shuttle Safety Assessment Report," issued prior to the first shuttle launch in 1981. Similar to the hazard analysis and tracking process introduced in Chapter 3, a hazard was classified by NASA as an acceptable risk only after

- All reasonable risk avoidance measures were identified, studied, and documented, and
- Project/program management made a decision to accept the risk on the basis of documented risk rationale.

Flying with "acceptable risks" was a standard part of NASA culture. At the time of the *Challenger* launch, the list of acceptable risks on the space shuttle filled six volumes.

7.2.2.2 Decision Making in the SRB Work Group

In an effort to identify problems in the design and performance of SRBs, engineers in the SRB Work Group used a Five-Step Decision Making Sequence:

1. Identify a signal of potential danger.
2. Officially acknowledge and escalate risk.
3. Review the evidence associated with the identified risk.
4. Official act indicating the normalization of deviance: accepted risk.
5. Shuttle launch.

In response to new signals of potential danger, the five-step decision-making process was repeated. Each time, evidence of joint performance initially interpreted as a deviation from expected performance was interpreted as within the bounds of "acceptable risk." The Work Group had, in a problem-driven, incremental fashion, developed a three-factor technical rationale to normalize deviance and deem a risk as an acceptable risk, consisting of

- A "safety margin"
- The "experience base"
- The "self-limiting" aspects of joint dynamics

Launch after launch, this repeated rationale convinced them that O-ring impingement erosion and blow-by were within the bounds of acceptable performance. In FRRs, the Work Group's official definition of the situation was conveyed up the hierarchy, creating at NASA a "uniform cultural construction of risk" for the booster joints. Essentially, because post-flight investigation showed that the secondary O-ring sealed and prevented an alarming amount of hot gases from escaping the SRB, a belief in the redundancy of the primary and secondary O-ring design was formed.

7.2.2.3 Construction of Risk

The formalization of the construction of risk began before the first shuttle flight in 1981. The SRBs had to be certified as flight-worthy through the assignment of Criticality categories on the Critical Items List (CIL). Criticality categories were formal labels assigned to each shuttle component, identifying the "failure consequences" should the component fail. The CIL entry for each item told why each component was an acceptable risk. The entries described the data and actions taken that the engineers believed precluded catastrophic failure. The "Rationale for Retention" included the "technical rationale," based on evidence from tests and flight experience that stated why the design should be retained for the critical item.

The NASA CIL identified C1 and C1R as follows:

- Criticality 1 (C1): Loss of life or vehicle if the component fails.
- Criticality 1R (C1R): Redundant components, the failure of both could cause loss of life or vehicle.

In November 1980, the SRB joint was classified as C1R, the "R" indicating the redundancy of the secondary O-ring in the field joint. Even though a new SRB Project Manager was named in November 1982 and he switched the CIL designation for the field joint O-ring problem from C1R to C1, Thiokol still considered the classification as a C1R. In addition, the C1 was waived for all subsequent flights. The waiver was found by the Presidential Commission to be nondeviant and within acceptable procedural guidelines of the existing NASA culture.

If a design change had been deemed necessary for the field joint O-ring design, the SRB production schedule would have to have been interrupted. To accomplish this, the engineers had to demonstrate that the proposed change would provide more than a marginal improvement in performance over the design currently in use because of the unknown safety hazards associated with the change. This was not deemed necessary by the SRB Work Group.

Also, because of the flight-after-flight belief in the redundancy of the primary and secondary O-ring design, the economic consequences of a mission loss were not considered when weighing short-term economic consequences of living with the O-ring problem because SRB joint problems were not defined as a threat to mission safety.

7.2.2.4 Three Safety Review Teams That NASA Utilized, Two Internal and One External

The Internal NASA Safety System consisted of the Safety, Reliability and Quality Assurance Program (SR & QA), and the Space Shuttle Crew Safety Panel (SSCSP). An external panel was added by Congress: the Aerospace Safety Advisory Panel (ASAP):

1. *Safety, Reliability and Quality Assurance Program:* The SR & QA consisted of an assembled team of Safety, Reliability, and Quality Assurance
   Engineers. The Safety Engineers were responsible for in-plant safety and
   flight safety. Reliability Engineers determined that contractor-supplied
   components and systems could be counted on to work as predicted. And
   Quality Assurance Engineers were responsible for the following procedural
   controls: the creation, assessment, and monitoring of hardware inspection
   programs, and identification and reporting of problems by monitoring computer problem-tracking systems.

   As the shuttle program matured from a developmental phase to an operational phase, the SR & QA depended on NASA for resources and legiti-
   macy, and both had been cut. Between 1970 and the *Challenger* disaster
   in 1986, NASA trimmed 71% of the SR & QA staff. At Marshall, SR &
   QA staff numbers were cut from 130 to 84. In 1982 when the shuttle was
   declared to be an operational vehicle, NASA either reorganized SR & QA
   offices or continued them with reduced personnel.

2. *Space Shuttle Crew Safety Panel:* The SSCSP was designed to function
   only during the shuttle's developmental period. From 1974 through 1979,
   it met twenty-six times; and although it addressed such issues as mission
   abort contingencies, crew escape systems, and acceptability of contractor-
   supplied equipment, O-ring problems were not discussed by the Panel.
   When the space shuttle officially became operational after its first four
   flights, the SSCSP was first combined with another Panel, and then eventually discontinued.

3. *Aerospace Safety Advisory Panel:* After the 1967 Apollo launch-pad fire
   that killed three astronauts, Congress added an external regulatory body
   composed of aerospace experts: the ASAP. This Panel did not assess booster
   joint design or in-flight anomalies of O-ring performance.

All three regulation bodies used a compliance strategy: the best strategy for accident
prevention. The goal of a compliance strategy is early detection and intervention,
negotiating to correct problems before they can cause any harm.

Because all three NASA safety process entities depended on the engineers in
the SRB Work Group for information, and the SRB Work Group had a prevailing
belief in the redundancy of the O-ring design, none of the safety panels identified the
O-ring design as a problem.

Although in her book, Dr. Vaughan proved that safety was not violated, the
role that safety played in the program severely diminished—not just organization-
ally, but also perceptually—over time. A post-tragedy analysis by the Presidential
Commission and the report of the House Committee found no evidence of the safety
team looking the other way, cooptation, hesitancy to speak, or willful concealment
on their part. They became "enculturated," that is, they followed the same incremental decision making of the SRB Work Group. They followed the same patterns
of information, decision streams, and commitment processes that the FRR rituals
generated; they also normalized deviance. It is speculation that even more safety

personnel participating in the same way on a daily basis would not have altered the scientific paradigm that prevailed in the years preceding the *Challenger* disaster.

7.2.2.5 Flight Readiness Review

The FRR was the final, formal review in an intricate process of launch preparation and decision making involving thousands of people and thousands of hours of engineering work. The goal of the FRR was to determine that the shuttle was ready to fly, and fly safely. Disputes between NASA and Thiokol engineers were commonplace before launches. At NASA, problems were the norm, and the word "anomaly" was part of everyday talk—to the point that disputes between contractor personnel and NASA officials did not contribute during the FRR to the escalation of concerns more than at any other time in the program.

7.2.2.6 The Night before the Launch

On the night before the launch, engineers at Morton Thiokol, the manufacturers of the shuttle's rockets, had recommended that the launch be delayed. Thiokol engineers knew that the next morning would have unprecedented low temperatures, to the point that the shuttle launch pad would be encased in ice and the temperature at liftoff would be just above freezing. In a late-night teleconference with NASA officials the night of January 27, 1986, Thiokol engineers, after conferring with their own management, changed their minds and agreed to a launch.

The norms, beliefs, and procedures that affirmed risk acceptability in the SRB Work Group were a part of the "worldview" that many engineers brought to the teleconference discussion. What had previously been an acceptable risk was no longer acceptable to many Thiokol engineers. The Five-Step decision sequence was initiated a final—and fatal—time, but with one alteration: it was followed not by a successful launch, but by a failure.

The original effort to delay the launch resulted from the fact that the coldest launch to date had been a January 1985 mission in which the calculated O-ring temperature was 53°F. After that flight, a post-flight inspection showed that a primary O-ring eroded to such an extent that it had not sealed at all, and hot propellant gases had penetrated the field joint, thus allowing the hot gases to "blow by" the primary toward the secondary O-ring.

In that instance, the secondary O-ring performed its intended function as a backup and successfully sealed the joint, but Thiokol engineers believed that more extensive blow-by could jeopardize the secondary O-ring, and thus mission safety.

In all the years of association between Thiokol and NASA officials at Marshall, this was the first time the contractor ever had come forward with a no-launch recommendation. NASA pointed out that, because no Launch Commit Criteria had ever been set for booster joint temperature, what Thiokol was proposing to do was create new Launch Commit Criteria on the eve of a launch. At one point, a Thiokol engineer was instructed by his own management to "take off his engineering hat and put on his management hat."

When Thiokol engineers were asked the rationale for agreeing to change their minds on the opinion not to launch unless the O-ring temperature was 53°F or greater, all said that they were influenced by facts not taken into account before their initial recommendation. These facts supported the consistent belief in the redundancy of the primary/secondary O-ring design that had carried through the SRB Work Group, launch after launch, despite the presence of blow-by. They believed that the secondary O-ring would seal the joint, supported by the following facts:

- Blow-by had been seen at both low and high temperatures. A previous shuttle flight that had suffered worse O-ring damage than the low-temperature flight of January 1985 had been launched in 75°F heat.
- In several tests where O-ring erosion was simulated by cutting parts of the O-ring and exposing the joint to high air pressure, results had shown that three times more erosion on the O-ring could occur and it would still seal the joint and work just fine, so the margin of safety was assumed to be large.
- During the initial pressurization at launch, if the primary O-ring experiences blow-by, there is only going to be so much blowing because it will seal in its seat. The joint is still going to seal and you are not going to have a catastrophic set of circumstances occur.

7.2.3 AFTERMATH

Going back to the two main questions asked by Dr. Vaughan:

1. Why, in the years preceding the *Challenger* launch, did NASA continue launching with a design known to be flawed?

   It was the belief in the redundancy of the primary/secondary O-ring design that was established by the SRB Work Group that allowed this design to continue to be flown. This was the case even with three safety review panels, and an escalation of concern that elevated the criticality of the O-ring problem from C1R to C1, at least by NASA.

2. Why did NASA launch the *Challenger* against the eve-of-launch objections by Thiokol engineers?

   In another display of bureaucratic accountability and structural secrecy, they (Thiokol engineers) deferred to others who had more authority to speak than they. As shown above, a belief in redundancy, and successful launch data in which that belief continued to result in successful flights, influenced those engineers who had concerns the night before the launch, to change their minds and not impose new criteria on the launch. In the rules-based culture in the Booster Working Group, Thiokol, and NASA overall, rigorous adherence to the engineering methods, routines, and lessons of the original technical culture from the development phase, and to the bureaucratic proceduralism of the organization, the whole shuttle system operated on the assumption that "deviation could be controlled but not eliminated," and no one believed anything other than what had held up so far: the belief in the redundancy of the design.

The Shuttle *Challenger* disaster represents a real-world example in which an organization, NASA, allowed the bridge between systems engineering and safety that had existed throughout NASA's history, from Mercury to Gemini to Apollo and on to the early development of the shuttle, to collapse. The Safety Breakdown Theory in this case was the result of several biases that contributed to the failure. As described in Chapter 6, The Glismann Effect—in the form of Pressure bias, Feedback bias, and Availability bias—was obviously present:

- *Pressure bias:* Congress and the White House established goals and made resource decisions that transformed the R&D space agency into a quasi-competitive business operation, complete with repeating production cycles, deadlines, and cost and efficiency goals. Safety personnel had been cut and resources had been reorganized. It had to have prevailed on the minds of NASA management that an omission in the President's State of the Union address would not have yielded precious political advantages.
- *Feedback bias:* The proper feedback—that the redundant design of the O-rings did not enforce the belief that they were able to control the release of hot propellant gases from the booster field joints—was not properly sent by lower-level engineers, nor received by responsible NASA stakeholders: The belief in the redundancy of the primary/secondary O-ring design supported the earlier claim that the shuttle was safe to fly, even as the Criticality was escalated by NASA from C1R to C1. Data regarding O-ring blow-by, erosion, and cold weather performance during previous shuttle launches were documented, but because these data did not support claims made earlier, they were not treated with a heightened level of safety concern.
- *Availability bias:* As stated in Chapter 6, because the frequency with which events come to mind is usually not an accurate reflection of their actual probability in reality, this leads to errors in judgment. With respect to the performance of the O-ring design launch after launch, Thiokol had data on redundancy and data on resiliency, but they had no data on how resiliency affected redundancy. After the tragedy, it took two non-engineer members of the Presidential Commission to put together two charts for the Commission Report to demonstrate how Thiokol might have arrayed O-ring performance versus ambient temperature data to create a strong signal: Of the flights launched above 65°F, three out of seventeen, or 17.6%, had anomalies. Of the flights launched below 65°F, 100% had anomalies. No Thiokol chart plotted ambient temperature against O-ring damage for each launch. Such data would have escalated the concern by showing the decrease in safe performance to the coupled decrease in O-ring resiliency with lower ambient temperature. However, the availability of the belief in the redundancy of the design supported the adage that "If you can think of it, it must be important."

A better bridge between safety and the systems engineering life cycle, not influenced by political pressure, the impression of a quasi-operational versus a developmental program, and reduced attention in the way of funding and resources, certainly would have helped to prevent this disaster.

7.3   NASA SHUTTLE *COLUMBIA* TRAGEDY

A visit to the Shuttle Pavilion at the USS Intrepid Museum on the Hudson River in New York City is an excellent opportunity to ogle and admire up close a great technical marvel of scientific achievement and American engineering superiority: the Shuttle Enterprise. Although it never flew as part of the shuttle fleet, it is an excellent tribute to NASA's success, and also a painful reminder of two of its worst failures: that of the Shuttle *Challenger* and the Shuttle *Columbia*.

In Flight STS-107 on February 1, 2003, the Shuttle *Columbia*, the first shuttle to orbit the Earth in 1981, disintegrated over Texas during re-entry slightly more than seventeen years to the day after the Shuttle *Challenger* tragedy. This again resulted in the death of all seven astronauts. The Shuttle Enterprise played a role in the Shuttle *Columbia* accident investigation. Figure 7.10 and Figure 7.11 show the location of the area of left wing damage that contributed to Shuttle *Columbia*'s demise.

7.3.1   INCIDENT

Approximately eighty-one seconds after liftoff on January 16, 2003, a briefcase-sized piece of foam that was suspected to contain ice somehow detached from the external fuel tank and impacted the left wing of the Shuttle *Columbia* main orbiter vehicle. This probably damaged the left wing enough to cause part of the shuttle's Thermal Protection System (TPS) tiles to be compromised. Upon re-entry into the Earth's atmosphere on February 1, 2003, the damage was magnified by the super-hot plasma, or ionized gas with an increased number of electrons, which conducted extremely high heat to the shuttle's surface, that the shuttle creates during re-entry. At one point, the left wing skin was pierced, allowing plasma to enter the wing. This seriously damaged the internal structure of the wing, and data show that wing sensors began to fail. *Columbia* eventually lost aerodynamic control at a speed near 20,000 kilometers per hour. The report of the Board assembled to investigate

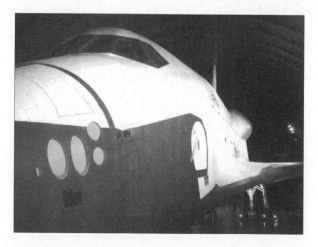

FIGURE 7.10   Shuttle *Columbia* left wing.

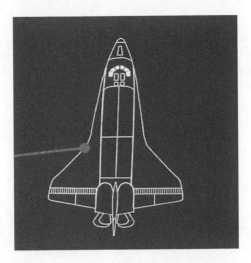

FIGURE 7.11   Wing damage location on Shuttle *Columbia*.

the tragedy, the *Columbia* Accident Investigation Board (or CAIB), referred to the failure as "main vehicle aerodynamic break-up."

7.3.2 ANALYSIS

As was the case with the O-ring erosion and blow-by as seen in the several flights before the Shuttle *Challenger* disaster, this flight was not the first time that foam had detached from the external fuel tank and caused damage to a shuttle. There had been numerous impacts on previous flights—by foam shrapnel much smaller than that which hit *Columbia*. Shuttles would typically return with hundreds of impacts larger than 2.5 centimeters. There were numerous reports in the Problem Reporting and Corrective Action (PRACA) decision-making information database system (referred to in the CAIB Report as a "dysfunctional database") that showed thermal tile damage.

Not only was foam impacting *not* considered in the design of the TPS, but also the high-level engineering requirements of the shuttle specifically state that nothing should have impacted the shuttle during launch. But foam had been hitting the shuttle consistently, and a team of chemists and engineers were working on the proper kind of application of foam to apply without using an older method that required environmentally damaging Freon®. The studies of foam impacts during shuttle flights all concluded that they did not pose a "flight safety" risk.

Because the main orbiters had returned from flight with no more than 2.5 centimeters of damage in some areas of the TPS, "foam-shedding" from the external fuel tank was considered a low-risk, anomalous event. It was unforeseen up to the launch of the *Columbia* in 2003 that a briefcase-sized piece of foam that may have contained a percentage of ice or "ablator" material (i.e., insulation that is placed under the tank's layer of foam to provide lightweight protection against high temperatures) would shed from the tank and hit one of the orbiter's most vulnerable areas: the leading edge of the wing, where the contour makes the placement of the TPS difficult. This view of an anomaly believed to be of little concern is another example of "normalized deviance" described by Dr. Vaughan: an instance in which an unprecedented anomaly is perceived to be routine.

The CAIB Report, released in August 2003, described many of the same organizational problems, and several new problems that did not exist up to the flight of the Shuttle *Challenger* in 1986. Although not described as such in the report, two things about NASA's funding, organization, and internal decision making leading up to the re-entry disaster of the Shuttle *Columbia* stick out:

1. *NASA's identity crisis:* The perception of the main orbiter as an operational, rather than a developmental, vehicle persisted in the wake of the Rogers Commission Report released after the Shuttle *Challenger* tragedy. By the time Flight STS-107 launched, the shuttle had become a bus to transfer crew and supplies to the International Space Station (ISS) and a repair vehicle for satellites. Also, budget, resource, and funding priorities shifted to the ISS, siphoning any ability of NASA to treat safety as an organizational priority with respect to shuttle flights.

   Finally, because either internal shuttle replacement vehicles had been proposed and subsequently taken off the table several times, or a privatization program for trips to the ISS were discussed but, at the time of Flight STS-107 had failed to come to fruition, emphasis on putting greater funds and personnel into the continued operation of the shuttle fleet did not occur.

2. *Post-launch denial:* After liftoff, several NASA employees and contractors had an opportunity to view video taken from a high-resolution camera mounted on a Cape Canaveral Air Force Station runway. One engineer said of the foam strike on the wing, "By far, it was the largest piece of debris we had ever seen come off the vehicle [External Fuel Tank] and the largest piece of debris we had seen impact the orbiter."

   A Debris Assessment Team had been assembled to assess the damage to the reinforced carbon-carbon (RCC) tiles. The team sent a request in order to obtain photos of the shuttle in orbit. This request was withdrawn by NASA management. The perception, as one engineer logged in his personal notebook: "Even if we saw something, we couldn't do anything about it." NASA management believed that it wanted to see the "worst-case" analysis results from the Debris Assessment Team without considering that the results might be contingent on reviews of photos, which might have shown the impact that the foam had on the wing. Because the impact happened at a vantage point not visible to the shuttle crew, only the photos might have allowed the team to assess actual damage.

   During the CAIB investigation, a lingering question was to find out what, if anything, NASA could have done to save the astronauts if the severity of the foam strike had been understood early on. It was determined that if the foam impact had damaged the orbiter and re-entry of the shuttle was indeed at risk, then there were only three mutually exclusive options available (i.e., if one decision was made, it would have excluded the other two as options):

a. Launch another shuttle as a rescue mission.
b. Perform a spacewalk repair mission.
c. Send the crew to the ISS to await a rescue vehicle.

None of these options were deemed viable. In one high-level NASA management conversation, one engineer said: "We don't have a [tile] repair material. We don't have the ability to get out and look at it. We don't have a [shuttle] robot arm. We don't have [spacewalking] tethers." As with the Shuttle *Challenger* disaster, an entrenched belief in the successful past performance of the shuttle despite evidence of anomalies, flight after flight, left a culture of denial that resulted in the assumption that nothing would go wrong.

7.3.3 AFTERMATH

Although the CAIB Report examined several budgetary and institutional "shortcomings," and flawed communication across NASA culture, it went so far as to describe the safety function as a "Broken Safety Culture." The investigation uncovered a troubling pattern in which Shuttle Program Management made erroneous assumptions about the robustness of a system based on prior success rather than on sound engineering practices, dependable engineering data, and rigorous testing. As an example, after the tragedy, one test conducted as part of the CAIB investigation was to shoot a chunk of external fuel tank foam at a shuttle wing mockup at the Southwest Research Institute in San Antonio, Texas, using a nitrogen gas cannon. This test allowed the Board to conclude: "without qualification, the foam did it." That test easily could have, and should have, been performed under an escalated safety investigation prior to the *Columbia* launch.

The Glismann Effect, in which a breakdown of safety occurred due to a confluence of biases, was evident in Pressure bias and Feedback bias. The Shuttle *Columbia* disaster represents a real-world example in which NASA, having received recommendations after the previous Shuttle *Challenger* disaster in the Rogers Report to "re-build" the bridge between systems engineering and safety, again resulted in a safety breakdown. The bridge, never fully repaired, collapsed yet again.

8 The Road Ahead

This book presented the tools of safety: identification and analysis of faults, failures, and hazards; the methodology of systems engineering; the life cycle as governed through technical and project processes; and the elements of management: culture, commitment, communication, and coordination. These three elements, or M – T – M, as described in Figure 1.1, are necessary to build the bridge that incorporates safety into the systems engineering life cycle. In Chapter 6, I described a theory of management breakdown, The Glismann Effect, in which bad decisions resulting from a confluence of biases might result in the discounting or exclusion of the importance of safety in the systems engineering life cycle. Chapter 7 presented real-world examples in which this exclusion led to disaster.

As a safety subject matter expert, systems engineer, or manager responsible for the incorporation of complex technical processes into a project or program life cycle, the question that must be asked is: Where do we go from here? This chapter proposes a road ahead. This road can be referred to as "The Three A's": Audit, Autonomy, and Authority.

1. *Audit:* In Chapter, 3, the audit roles and responsibilities of the Independent Safety Assessor (ISA) were discussed in terms of collaboration with and information presented to the stakeholders. Along with safety audits performed by other interested parties throughout the system life cycle, those conducted by the ISA must have the highest visibility to top management, in terms of open issues, lessons learned, and the roadmap to resolve concerns. Safety subject matter experts should be part of the ISA auditing team, and contribute to the presentation of the audit findings to top management.
2. *Autonomy:* To avoid the biases that contribute to the safety breakdown discussed in Chapter 6, there should be a mechanism of empowerment for systems engineers and safety professionals to be able to bypass the bureaucracy and hierarchy, especially early in the life cycle, in order to properly escalate safety concerns. This is particularly crucial in environments in which an increasing level of risk is accepted as the life cycle matures, based on the fact that failures did not occur while these risks existed.
3. *Authority:* On the road ahead, it must be the responsibility of top management in complex technical organizations to grant authority to the systems engineers and safety subject matter experts responsible for establishing and maintaining safety throughout the system life cycle. In her book *Engineering a Safer World,* Professor Nancy Leveson of MIT describes a "Safety Culture": a subset of culture that reflects the general attitude and approaches to safety and risk assessment. She states that "the struggle for a good safety culture will never end because it must continually fight against

the functional pressures of the work environment." Management must secure the funding and provide the political support to instill and maintain this safety culture. This point is also emphasized in the findings presented in the CAIB Report, Volume I, Chapter 7: The Accident's Organizational Causes:

"Organizations that successfully deal with high-risk technologies create and sustain a disciplined safety system capable of identifying, analyzing, and controlling hazards throughout a technology's life cycle. The practices noted here suggest that responsibility and authority for decisions involving technical requirements and safety should rest with an independent technical authority."

Instead of an environment that leads to a breakdown of the importance of safety in a systems engineering methodology, management must establish feedback loops of audits and performance assessments, reporting systems, and anomaly, incident, and accident investigation. The term "lessons learned" must be part of the tailoring process. Lessons learned must help to escalate identified safety issues or incidents from a rote exercise to an opportunity to teach a method of systemic and programmatic improvement throughout a life cycle.

In the future, I hope to secure funding, either through grants or fellowships, to conduct studies to further investigate the validity of The Glismann Effect. I would like to investigate more real-world examples, conduct interviews, and apply techniques of quantification to map the claim that biases lead to disasters. I would also like to further investigate methods of preventing these occurrences.

It is my hope for all the interested stakeholders who have read this book that the information presented herein will prove useful on the road ahead. I hope that all stakeholders will utilize M – T – M, and put the necessary time and effort into selling the concept of excellence. The sooner that safety is incorporated into the systems engineering life cycle and the earlier that the Culture, Commitment, Communication, and Coordination are employed to build the bridge, the greater the chance of providing opportunities for success instead of paths to disaster. The more that all stakeholders and team members are excited, enthusiastic, and empowered, the more they will press on the road to success: to build the bridge between systems engineering and safety, understand the big picture, have and share the vision for success, and work to achieve the goal.

Bibliography

Badiru, Adedji B., *Triple C Model of Project Management Communication, Cooperation, and Coordination,* Taylor & Francis Group / CRC Press, Boca Raton, FL, 2008.

Blanchard, Benjamin S., *Systems Engineering Management,* Wiley-Interscience, New York, 1991.

Cabbage, Michael, and Harwood, William, *Comm Check The Final Flight of Shuttle Columbia,* Free Press, New York, 2004.

Columbia Accident Investigation Board, Final Report, August 1993.

Diehl, Alan E., *Silent Knights: Blowing the Whistle on Military Accidents and Their Cover-Ups,* Brasseys, Inc., Dulles, VA, 2002.

Englund, Randall, and Bucero, Alfonso, *Project Sponsorship Achieving Management Commitment for Project Success,* Jossey-Bass, New York, 2006.

Evans, Dylan, How to Beat the Odds at Judging Risk, *Wall Street Journal,* May 14, 2012.

Galvin, Molly, PE perseveres, 10 years after Challenger explosion, *Engineering Times,* National Society of Professional Engineers, August 1995.

Hall, Joseph Lorenzo, Columbia and Challenger: Organizational failure at NASA, *Space Policy,* 19, 239–247, 2003.

INCOSE, Haskins, Celia (Ed.), *International Council on Systems Engineering (INCOSE), Systems Engineering Handbook: A Guide for System Life Cycle Processes and Activities,* Version 3.2.1, INCOSE, San Diego, CA, January 2011.

Jackson, Lisa, and Schmidt, Gerry, *Transforming Corporate Culture 9 Natural Truths for Being Fit to Compete,* Evolved Leaders Publishing, Littleton, CO, 2011.

Janis, Irving L., *Groupthink: Psychological Studies of Policy Decisions and Fiascoes, second edition,* Houghton Mifflin, Boston, MA, 1982.

Kahneman, Daniel, *Thinking, Fast and Slow,* Farrar, Straus and Giroux, New York, 2011.

Leveson, Nancy, *Engineering a Safer World Systems Thinking Applied to Safety,* The MIT Press, Cambridge, MA, 2012.

MIL-HDBK-338B Electronic Reliability Design Handbook 1, October 1998.

MIL-STD-1629A: Procedures for Performing a Failure Mode. Effective and Critical Analysis, 24 November 1980.

MIL-STD-882C: Military Standard System Safety Program Requirements, 19 January 1993.

National Aeronautics and Space Administration (NASA), Methodology for Conduct of Space Shuttle Program Hazard Analyses, NSTS 22254 Revision B, December 1993.

National Aeronautics and Space Administration (NASA), *Safety, Reliability, Maintainability and Quality Provisions for the Space Shuttle Program,* NSTS 5300.4, September 1997.

National Research Council, *Preparing for the High Frontier, The Role and Training of NASA Astronauts in the Post-Space Shuttle Era,* The National Academies Press, Washington, DC, 2011.

NIST, A Conceptual Framework for Systems Fault Tolerance, NIST.gov.

O'Rourke IV, James S., *Management Communication: A Case-Analysis Approach, third edition,* Pearson Prentice Hall, Upper Saddle River, NJ, 2007.

Perrow, Charles, *Normal Accidents Living with High-Risk Technologies,* Princeton University Press, Princeton, NJ, 1999.

Schwoebel, Richard L., *Explosion Aboard the Iowa,* Naval Institute Press, Annapolis, MD, 1999.

US Department of Defense, MIL-STD-882C Military Standard System Safety Program Requirements, US Department of Defense, January 1993.

US Department of Defense, MIL-STD-1629A Military Standard Procedures for Performing a Failure Mode, Effects and Criticality Analysis, US Department of Defense, November 1980.

US Department of Energy (DOE), *Human Performance Improvement Handbook,* DOE-HDBK-1028-2009, January 2009.

US Department of Transportation, Federal Railroad Administration, Standards for Processor-Based Signal and Train Control Systems, Subpart H, 70 FR 11095, March 2005.

US Department of Transportation, Federal Transit Administration, Hazard Analysis Guidelines for Transit Projects, DOT-FTA-MA-26-5005-00-01 1/2000.

US Department of Transportation, Federal Transit Administration, Handbook for Transit Safety and Security Certification, FTA-MA-90-5006-02-01 11/2002.

Vaughan, Diane, *The Challenger Launch Decision: Risky Technology, Culture and Deviance at NASA,* University of Chicago Press, Chicago, 1996.

Vincoli, Jeffrey W., *Basic Guide to System Safety, second edition,* CSP Wiley, New York, 2006.

Wilde, Gerald J.S., *Target Risk 2: A New Psychology of Safety and Health,* PDE Publications, Toronto, Canada, 2001.

Wilensky, Harold, *Organizational Intelligence,* Basic Books, Inc., New York, 1967.

Willcoxson, Lesley, and Millet, Bruce, The management of organisational culture, *Australian Journal of Management & Organisational Behaviour,* 3(2), 91–99, 2000.

Index